Springer Monographs in Mathematics

This series publishes advanced monographs giving well-written presentations of the "state-of-the-art" in fields of mathematical research that have acquired the maturity needed for such a treatment. They are sufficiently self-contained to be accessible to more than just the intimate specialists of the subject, and sufficiently comprehensive to remain valuable references for many years. Besides the current state of knowledge in its field, an SMM volume should ideally describe its relevance to and interaction with neighbouring fields of mathematics, and give pointers to future directions of research.

More information about this series at http://www.springer.com/series/3733

Élisabeth Gassiat

Universal Coding and Order Identification by Model Selection Methods

 Springer

Élisabeth Gassiat
Laboratoire de Mathématiques
Université Paris-Sud
Orsay Cedex, France

Translated by Anna Ben-Hamou, LPSM, Sorbonne Université, Paris, France

ISSN 1439-7382   ISSN 2196-9922 (electronic)
Springer Monographs in Mathematics
ISBN 978-3-030-07167-7   ISBN 978-3-319-96262-7 (eBook)
https://doi.org/10.1007/978-3-319-96262-7

Mathematics Subject Classification (2010): 68P30, 62C10

This Springer imprint is published by the registered company Springer Nature Switzerland AG
The registered company address is: Gewerbestrasse 11, 6330 Cham, Switzerland

TO HOW, WHO LIKE BOOKS IN VARIOUS (CODING) LANGUAGES!

Preface

Quantifying information contained in a set of messages is the starting point of information theory. Extracting information from a dataset is at the heart of statistics. Information theory and statistics are thus naturally linked together, and this course lies at their interface.

The theoretical concept of information was introduced in the context of research on telecommunication systems. The basic objective of information theory is to transmit messages in the most secure and least costly way. Messages are encoded, then transmitted, and finally decoded at reception. Those three steps will not be investigated here. We will essentially be interested in the first one, the coding step, in its multiple links with statistical theory, and in the rich ideas which are exchanged between information theory and statistics. The reader who is interested in a more complete view of the basic results in information theory can refer, for instance, to the two (very different) books: [1, 2].

This book is mostly concerned with lossless coding, where the goal is to encode in a deterministic and decodable way a sequence of symbols, in the most efficient way possible, in the sense of the codewords' length. The gain of a coding scheme is measured through the compression rate, which is the ratio between the codeword's length and the coded word's length. If the sequence of symbols to be encoded is generated by a stochastic process, a coding scheme will perform better if more frequent symbols are encoded with shorter codewords. This is where statistics comes into play: if one only has incomplete knowledge of the underlying process generating the sequence of symbols to be encoded, then, in order to improve the performance of the coding compression, one had better use what can be inferred about the law of the process from the first observed symbols. Shannon's entropy is the basic quantity of information allowing one to analyze the compression performance of a given coding method. When possible, one defines the entropy rate of a process as the limit, as n tends to $+\infty$, of the Shannon's entropy of the law of the first n symbols, normalized by n.

In the first chapter, we will see that the asymptotic compression rate is lower bounded by the entropy rate of the process' distribution producing the text to encode, provided that this process is ergodic and stationary. We will also see that every coding method can be associated with a probability distribution in such a way that the compression performance of a code associated with a distribution Q for a process with distribution P is given by an information divergence between P and Q. The setting is laid down: the problem of universal coding is to find a coding method (hence a sequence of distributions Q_n) which asymptotically realizes (when the number n of symbols to be encoded tends to infinity) the optimal compression rate, for the largest possible class of distributions P. While investigating this question, we will particularly interested in understanding the existing links between universal coding and statistical estimation, at all levels, from methods and ideas to proofs.

We will see, in Chap. 2, that in the case of a sequence of symbols with values in a finite alphabet, it is possible to find universal coding methods for the class of all distributions of stationary ergodic processes. Before studying statistical methods in a strict sense, we will present Lempel–Ziv coding, which relies on the simple idea that a codeword's length can be shortened by taking advantage of repetitions in the word to be encoded. We will then present different quantification criteria for compression capacities and will see that those criteria are directly related to well-known statistical methods: maximum likelihood estimation and Bayesian estimation. We will take advantage of the approximation of stationary ergodic processes by Markov chains with arbitrary memory. Such chains are called context tree sources in information theory, and variable length Markov chains in statistical modelization. Few things are known for nonparametric classes, even in finite alphabets. We will present the example of renewal processes for which we will see that the approximation by variable length Markov chains is a good approximation.

Chapter 3 then tackles the problem of coding over infinite alphabets. When trying to encode sequences of symbols with values in a very large alphabet (which may then be seen as infinite), one encounters various unsolved problems. In particular, there is no universal code. One is then confronted with problems related to model selection and order identification. After having laid down some milestones (coding of integers, necessary and sufficient conditions for the existence of a weakly universal code over a class of process distributions), we study more particularly, as a first attempt toward a better understanding of these questions, classes of process distributions corresponding to sequences of independent and identically distributed variables, characterized by the speed of decrease at infinity of the probability measure. An alternative idea is to encode the sequence of symbols in two steps: first the pattern (how repetitions are arranged), then the dictionary (the letters used, in their order of appearance). We will see that the information contained in the message shape, measured by the entropy rate, is the same as that contained in the whole message. However, although it is not possible to design a universal code for the class of memoryless sources (sequences of independent and identically random variables) with values in an infinite alphabet, it is possible to obtain a universal code for their pattern.

Chapter 4 deals with the question of order identification in statistics. This is a model selection problem which arises in various practical situations and which aims at identifying an integer characterizing the model: length of dependency for a

Markov chain, number of hidden states for a hidden Markov chain, number of populations in a population mixture. The coding ideas and techniques presented in the previous chapter have recently led to some new results, in particular concerning latent variable models such as hidden Markov models. We finally show that the question of order identification relies on a delicate understanding of likelihood ratio trajectories. We point out how this can be done in the case of population mixtures.

At the end of each chapter, one may find bibliographical comments, important references, and some open problems.

The original French version of this book [3] resulted from the editing of lecture notes intended for students in *Master 2 of Probability and Statistics* and doctoral students at Orsay University (Paris-Sud). It is accessible to anyone with a graduate level in mathematics, with basic knowledge in mathematical statistics. The only difference between this translated version and the first edition is in the remark following Theorem 3.6, where we mention some progress that has been made since then.

Except in Chap. 4, all the proofs are detailed. I chose to recall all the necessary tools needed to understand the text, usually by giving a detailed explanation, or at least by providing a reference. However, the last part of Chap. 4 contains more difficult results, for which the essential ideas are given but the proofs are only sketched.

I would like to thank everyone who have read this book when it was in progress. It has greatly benefited from their attention. I am particularly grateful to Gilles Stoltz and to his demanding reading, and to Grégory Miermont for his canonical patience.

Orsay, France   Élisabeth Gassiat

References

1. T.M. Cover, J.A. Thomas, *Elements of Information Theory*. Wiley Series in Telecommunications (Wiley and Sons, New York, 1991)
2. I. Csiszár, J. Korner, *Information Theory: Coding Theorems for Discrete Memoryless Systems*, 3rd edn. (Akademia Kiado, Budapest, 1981)
3. É. Gassiat, *Codage universel et identification d'ordre par sélection de modéles* (Société mathématique de France, 2014)

Contents

Notations

$\mathscr{X}$	A finite or countable set called an alphabet, except in Sects. 2.2.3, 2.3.1, and 4.3, where $\mathscr{X}$ is a complete separable metric space
$\mathscr{X}^*$	$\bigcup_{n \geq 1} \mathscr{X}^n$, called the set of words (with letters in the alphabet $\mathscr{X}$)
$x_{1:n}$	The n-tuple $(x_1, \ldots, x_n)$
P	A probability (over a subset of $\mathscr{X}$ or of $\mathscr{X}^*$)
$\mathbb{P}$	The distribution of a sequence of random variables $(X_n)_{n \in \mathbb{N}}$ or $(X_n)_{n \in \mathbb{Z}}$
$\mathbb{P}_n$	The distribution of $X_{1:n}$ if $(X_n)_{n \in \mathbb{N}}$ has distribution $\mathbb{P}$
$\mathbb{P}^k$	The k-Markovian approximation of $\mathbb{P}$
$H(P)$ or $H(X)$	The entropy of P (or of the random variable X with distribution P)
$H_*(\mathbb{P})$	The entropy rate of $\mathbb{P}$
$D(P\|Q)$	The relative entropy, or Kullback information of P with respect to Q
$I(X; Y)$	The mutual information of random variables X and Y
$\mathscr{E}(n)$	The Elias code of an integer n
$\overline{R}_n(\mathscr{C})$	The minimax redundancy over class $\mathscr{C}$
$\underline{R}_n(\mathscr{C})$	The Bayesian redundancy over class $\mathscr{C}$
$R_n^*(\mathscr{C})$	The regret over class $\mathscr{C}$
NML	The normalized maximum likelihood probability
$\Gamma(.)$	The Gamma function
$\mathscr{L}_\ell$	The simplex of $\mathbb{R}^\ell$
$\mathbb{S}_\ell$	The set of $(\ell - 1)$-tuples $(x_1, \ldots, x_{\ell-1})$ such that, letting $x_\ell = 1 - x_1 - \ldots - x_{\ell-1}$, we have $x_{1:\ell} \in \mathscr{L}_\ell$
KT	The Krichevsky–Trofimov distribution
CTW	The double mixture probability called Context Tree Weighting
Λ_f	The set of memoryless sources with marginal dominated by f

$\ell_n(\cdot)$	Log-likelihood
$v_n(\cdot)$	An empirical process
$h(\cdot,\cdot)$	Hellinger distance
$N(\mathscr{F},\varepsilon)$	The minimum number of brackets of size ε covering $\mathscr{F} \subset L^2(f^*\mathrm{d}\mu)$

Abstract

The purpose of these notes is to highlight the deep connections between Information Theory and Statistics. Indeed, universal coding and adaptive compression are strongly linked to statistical inference methods for random processes, such as maximum likelihood or Bayesian techniques. We first introduce classical tools for coding on finite alphabets, then we present the recent theory of universal coding on infinite alphabets. We show how it allows us to solve order identification problems, in particular for hidden Markov models.

Chapter 1
Lossless Coding

Abstract The goal here is to encode a sequence of symbols in such a way that it is possible to decode it perfectly (lossless coding), and to decode it sequentially (prefix coding). One may then relate codes and probabilities: this is the essence of the Kraft-McMillan inequalities. If one aims at minimizing the codeword's length, Shannon's entropy gives an intrinsic limit, when the word to be encoded is regarded as a random variable. When the distribution of this random variable is known, then the optimal compression rate can be achieved (Shannon's coding and Huffman's coding). Moreover, as codeword lengths are identified with probability distributions, for any probability distribution, one may design a prefix code which encodes sequentially. This will be referred to as "coding according to this distribution". Arithmetic coding, based on a probability distribution which is not necessarily the one of the source, will be particularly detailed. In this way, the algorithmic aspect of coding and the modeling of the source distribution are separated. Here the word "source" is used as a synonym for a random process. We finally point out some essential tools needed to quantify information, in particular the entropy rate of a process. This rate appears as an intrinsic lower bound for the asymptotic compression rate, for almost every source trajectory, as soon as it is ergodic and stationary. This also shows that is is crucial to encode words in blocks. Arithmetic coding has the advantage of encoding in blocks and "online". If arithmetic coding is devised with the source distribution, then it asymptotically achieves the optimal compression rate. In the following chapters, we will be interested in the question of adapting the code to an unknown source distribution, which corresponds to a fundamentally statistical question.

Let $\mathscr{X}$ be a set, which, throughout this chapter, is assumed to be finite or countable, except in the case of Huffman coding, for which $\mathscr{X}$ is assumed to be finite. The set $\mathscr{X}$ is called the *alphabet*, its elements are called *letters*, and finite sequences of letters are called *words*. The set of words, i.e. of finite sequences of elements of $\mathscr{X}$, will be denoted by $\mathscr{X}^*$:

$$\mathscr{X}^* = \bigcup_{n=1}^{\infty} \mathscr{X}^n.$$

© Springer International Publishing AG, part of Springer Nature 2018
É. Gassiat, *Universal Coding and Order Identification by Model
Selection Methods*, Springer Monographs in Mathematics,
https://doi.org/10.1007/978-3-319-96262-7_1

If n is an integer and $x_1, \ldots, x_n$ are letters, the word $(x_1, \ldots, x_n)$ will be denoted by $x_{1:n}$. A word $x \in \mathscr{X}^n$ is of *length n*, which will be denoted by $\ell(x) = n$. We denote by "." concatenation operation between words. Also, we introduce an additional word, denoted by $\emptyset$, satisfying $\ell(\emptyset) = 0$ and $x \cdot \emptyset = x$. This word, called the *empty word*, will not be encoded.

We will encode words of $\mathscr{X}^*$ by words of $\mathscr{Y}^*$, where $\mathscr{Y}$ is a finite alphabet. The code should be decodable and as short as possible. Although all our results could be written for an arbitrary finite $\mathscr{Y}$, we fix from now on

$$\mathscr{Y} = \{0, 1\}.$$

Encoded words are thus written in binary language, their letters are sometimes referred to as *bits*, and all logarithms used in this book are base 2 (except in Sect. 4.3). By convention

$$0 \log_2 0 = 0.$$

Let A be a subset of $\mathscr{X}^*$.

- A *lossless code* is an injection $f : A \to \mathscr{Y}^*$. There exists a function ϕ such that $\phi \circ f = \mathrm{Id}_A$, the identity function in A. This function ϕ is the *decoding function*: if one encodes $x_{1:n}$ by $f(x_{1:n}) = y_{1:m}$, then one retrieves the coded word through $\phi(y_{1:m}) = x_{1:n}$.
- The *compression rate* is the ratio between the length of the codeword and the length of the initial word:
$$\frac{\ell[f(x)]}{\ell(x)}.$$

- A code is said to be *uniquely decodable* when one does not need to separate words to decode: if $\omega_1, \ldots, \omega_n, \omega'_1, \ldots, \omega'_m$, are words of A, then

$$f(\omega_1) \ldots f(\omega_2) \ldots f(\omega_n) = f(\omega'_1) \ldots f(\omega'_2) \ldots f(\omega'_m)$$
$$\implies \quad n = m \text{ and } \omega_1 = \omega'_1, \ldots, \omega_n = \omega'_n.$$

- A code is said to be *prefix* when no codeword is the prefix of another codeword: if ω and ω' are words of A, then

$$f(\omega) \cdot y = f(\omega') \quad \implies \quad y = \emptyset \text{ and } \omega = \omega'.$$

This means that, when we arrive at the end of a codeword, we know it: this is sometimes referred to as an *instantaneously decodable code*.

If a code is prefix, then it is uniquely decodable.

1.1 Kraft-McMillan Inequalities

Kraft-McMillan Inequalities allow one to identify a code length with a sub-probability. Theorem 1.1 states that the length function of a uniquely decodable code is equal to minus the logarithm, in base 2, of a sub-probability, and the reverse, Theorem 1.2, states that if an integer-valued function is minus the logarithm, in base 2, of a sub-probability, then there exists a prefix code of which it is the length function.

Theorem 1.1 *Let $A \subset \mathscr{X}^*$ be a set of words. If the function $f : A \to \mathscr{Y}^*$ is uniquely decodable, then*

$$\sum_{\omega \in A} 2^{-\ell[f(\omega)]} \leqslant 1.$$

Proof Let us first consider the case where $\mathscr{X}$ is finite. Let m be any fixed integer and

$$p = \max \left\{ \ell[f(\omega)] \ : \ \omega \in A, \ \ell(\omega) \leqslant m \right\}.$$

As the set $\{\omega \in A, \ \ell(\omega) \leqslant m\}$ is finite, p is a finite integer. Thus

$$\left\{ \sum_{\ell(\omega) \leqslant m} 2^{-\ell[f(\omega)]} \right\}^n = \sum_{\substack{(\omega_1, \dots, \omega_n) \in A^n \\ \ell(\omega_1) \leqslant m, \dots, \ell(\omega_n) \leqslant m}} 2^{-\sum_{i=1}^n \ell[f(\omega_i)]}$$

$$\leqslant \sum_{h \leqslant np} \sum_{\substack{(\omega_1, \dots, \omega_n) \in A^n \\ \ell(f(\omega_1)) + \dots + \ell(f(\omega_n)) = h}} 2^{-h}.$$

As f is uniquely decodable, there are no two distinct sequences of words $\omega_1, \dots, \omega_n$ yielding the same concatenated sequence $f(\omega_1) \dots f(\omega_n)$ and each word of $\mathscr{Y}^h$ corresponds to at most one sequence $\omega_1, \dots, \omega_n$. We thus have

$$\left\{ \sum_{\ell(\omega) \leqslant m} 2^{-\ell[f(\omega)]} \right\}^n \leqslant \sum_{h \leqslant np} 2^h \cdot 2^{-h} = np$$

and for all n,

$$\sum_{\ell(\omega) \leqslant m} 2^{-\ell[f(\omega)]} \leqslant (np)^{1/n}.$$

Taking the limit as n tends to infinity, we get $\sum_{\ell(\omega) \leqslant m} 2^{-\ell[f(\omega)]} \leqslant 1$. This being true for all m, the inequality extends to the sum over all finite words.

Let us now consider the general case where $\mathscr{X}$ is countable. For all finite subsets B of A, the function $f : B \to \mathscr{Y}^*$ is uniquely decodable, which implies, by the above result,

$$\sum_{\omega \in B} 2^{-\ell[f(\omega)]} \leqslant 1.$$

The family of positive real numbers $(2^{-\ell[f(\omega)]})_{\omega \in A}$ is summable and its sum, which is the supremum of $\sum_{\omega \in B} 2^{-\ell[f(\omega)]}$ for B a finite subset of A, is less than or equal to 1.

Theorem 1.2 *If λ is a function from a subset A of $\mathscr{X}^*$ with integer values, satisfying*

$$\sum_{\omega \in A} 2^{-\lambda(\omega)} \leqslant 1,$$

then there exists a prefix code from A to $\mathscr{Y}^*$ such that, for all $\omega \in A$, $\ell[f(\omega)] = \lambda(\omega)$.

Proof (Proof of Theorem 1.2) In the case where A is finite, the proof relies on the identification between a prefix code and a binary tree. By a *binary tree*, we mean a rooted tree in which each node has at most two descendants. Nodes with no descendant are called *leaves* of the tree. By convention, we may choose the left descendant to correspond to 0 and the right descendant to correspond to 1, so that leaves of the tree are words composed of 0's and 1's. A binary tree is then identified with a subset of $\mathscr{Y}^*$. Here are two examples of binary trees:

Tree 1 is *complete*, in the sense that each node has 0 or 2 descendants. Tree 2 is not complete.

Each node (either a leaf or an *internal node*) is associated to a word.

The *depth* of a node is the length of the word associated to it.

Nodes of depth 1 are then associated to 0 or 1, and the number of possible nodes at depth p is 2^p. If the tree has n leaves with depths $p_1, \ldots, p_n$, respectively, we have $2^{-p_1} + \cdots + 2^{-p_n} = 1$ if the tree is complete, and $2^{-p_1} + \cdots + 2^{-p_n} \leqslant 1$ in any case.

The inequality of Theorem 1.2 thus means that one can construct a binary tree in such a way that each $\omega \in A$ is associated (bi-uniquely) to a leaf of depth $\lambda(\omega)$, and the tree is complete if and only if $\sum_{\omega \in A} 2^{-\lambda(\omega)} = 1$.

Denote by $f(\omega)$ the leaf (identified with a word of $\mathscr{Y}^*$) associated to ω. This clearly defines a prefix code, because each word which is the prefix of a word associated to a leaf of the tree is associated to an internal node.

Let us now write the general proof. Let us order the set $A = \{\omega_1, \omega_2, \ldots\}$ by increasing order in λ: if $\lambda(\omega_i) < \lambda(\omega_j)$, then $i < j$. We define the encoding $f(\omega_i)$ of ω_i as being the first $\lambda(\omega_i)$ numbers in the binary expansion of

$$z_i = \sum_{j < i} 2^{-\lambda(\omega_j)},$$

(an empty sum is set to null, so that $z_1 = 0$). In other words, if

$$z_i = \sum_{k \geqslant 1} a_k(z_i) 2^{-k},$$

with $a_k(z_i) \in \{0, 1\}$ and $k \geqslant 1$, then

$$f(\omega_i) = a_1(z_i) \ldots a_{\lambda(\omega_i)}(z_i).$$

If f is not a prefix code, then there exists a prefix $f(\omega_i)$ of $f(\omega_{i'})$ with $i \neq i'$. Either $\lambda(\omega_i) = \lambda(\omega_{i'})$, i.e. $f(\omega_i)$ and $f(\omega_{i'})$ have the same length and are equal, in which case we may choose the order such that $i < i'$, or $\lambda(\omega_i) < \lambda(\omega_{i'})$ in which case $i < i'$. In all cases, we may assume $i < i'$. We have

$$z_i = \sum_{j<i} 2^{-\lambda(\omega_j)} = \sum_{k \geqslant 1} a_k(z_i) 2^{-k},$$

and this expansion is finite: $a_k(z_i)$ is zero for k large enough. Similarly

$$z_{i'} = \sum_{j<i'} 2^{-\lambda(\omega_j)} = \sum_{k \geqslant 1} a_k(z_{i'}) 2^{-k},$$

and this expansion is finite. We have

$$z_{i'} - z_i = \sum_{i \leqslant j < i'} 2^{-\lambda(\omega_j)}.$$

Also, as $f(\omega_i)$ is a prefix of $f(\omega_{i'})$, $a_k(z_i) = a_k(z_{i'})$ for $k = 1, \ldots, \lambda(\omega_i)$. We thus have

$$z_{i'} - z_i \leqslant \sum_{k > \lambda(\omega_i)} a_k(z_{i'}) 2^{-k},$$

which implies $2^{-\lambda(\omega_i)} \leqslant \sum_{k > \lambda(\omega_i)} a_k(z_{i'}) 2^{-k}$. But this is impossible because $a_k(z_{i'})$ is zero for k large enough.

Example 1.1 It is impossible to encode, in a uniquely decodable way, four words $\omega_1, \ldots, \omega_4$ with words of length $\lambda(\omega_1) = 1$, $\lambda(\omega_2) = 2$, $\lambda(\omega_3) = 2$ and $\lambda(\omega_4) = 3$. Indeed,

$$\frac{1}{2} + \frac{1}{2^2} + \frac{1}{2^2} + \frac{1}{2^3} = 1 + \frac{1}{2^3} > 1.$$

- It is possible to encode, in a uniquely decodable way, four words $\omega_1, \ldots, \omega_4$ with words of length $\lambda(\omega_1) = 1$, $\lambda(\omega_2) = 2$, $\lambda(\omega_3) = 3$ and $\lambda(\omega_4) = 3$. Indeed,

$$\frac{1}{2} + \frac{1}{2^2} + \frac{1}{2^3} + \frac{1}{2^3} = 1.$$

$z_1 = 0$ and $\lambda(\omega_1) = 1$, thus $f(\omega_1) = 0$. $z_2 = \frac{1}{2}$ and $\lambda(\omega_2) = 2$, so that $f(\omega_2) = 10$. $z_3 = \frac{1}{2} + \frac{1}{2^2}$ and $\lambda(\omega_3) = 3$, so that $f(\omega_3) = 110$. $z_4 = \frac{1}{2} + \frac{1}{2^2} + \frac{1}{2^3}$ and $\lambda(\omega_4) = 3$, so that $f(\omega_3) = 111$. This code is identified with the binary Tree 1 of Fig. 1.1.

- It is possible to encode, in a uniquely decodable way, five words $\omega_1, \ldots, \omega_5$ with words of length $\lambda(\omega_1) = 2$, $\lambda(\omega_2) = 3$, $\lambda(\omega_3) = 3$, $\lambda(\omega_4) = 5$ and $\lambda(\omega_5) = 7$.

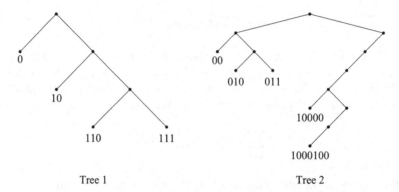

Fig. 1.1 Two binary trees

Indeed,

$$\frac{1}{2^2} + \frac{1}{2^3} + \frac{1}{2^3} + \frac{1}{2^5} + \frac{1}{2^7} = \frac{1}{2} + \frac{1}{2^5} + \frac{1}{2^7} \leqslant 1.$$

$z_1 = 0$ and $\lambda(\omega_1) = 2$, thus $f(\omega_1) = 00$. $z_2 = \frac{1}{2^2}$ and $\lambda(\omega_2) = 3$, thus $f(\omega_2) = 010$. $z_3 = \frac{1}{2^2} + \frac{1}{2^3}$ and $\lambda(\omega_3) = 3$, thus $f(\omega_3) = 011$. $z_4 = \frac{1}{2^2} + \frac{1}{2^3} + \frac{1}{2^3} = \frac{1}{2}$ and $\lambda(\omega_4) = 5$, thus $f(\omega_4) = 10000$. $z_5 = \frac{1}{2^2} + \frac{1}{2^3} + \frac{1}{2^3} + \frac{1}{2^5} = \frac{1}{2} + \frac{1}{2^5}$ and $\lambda(\omega_4) = 7$, thus $f(\omega_5) = 1000100$. This code is identified with the binary Tree 2 of Fig. 1.1.

Remarks 1.2 The identification between a tree and a prefix code gives another proof of Theorem 1.1 if the encoding is assumed to be prefix and the set A finite.

- Theorem 1.2 shows that, as far as compression is concerned, there is no advantage to using uniquely decodable codes which are not prefix. There exist uniquely decodable codes which are not prefix, but for such codes, one may devise a prefix code with the same length function.
- If we define, for all $\omega \in A$, $Q(\omega) = 2^{-\ell[f(\omega)]}$, we then have $\sum_{\omega \in A} Q(\omega) \leqslant 1$. Adding an element e in the set of words to be encoded, Q may be completed to a probability: set $A' = A \cup \{e\}$, and $Q(e) = 1 - \sum_{\omega \in A} Q(\omega)$. The length of the word which encodes ω is identified with minus the logarithm, base 2, of $Q(\omega)$ where Q is a probability over the set of words being encoded. This is referred to as coding *according to the distribution Q*.

1.2 Quantifying Information

If $X_1, \ldots, X_n$ are n random variables, we denote by $X_{1:n}$ the n-tuple $(X_1, \ldots, X_n)$.

We introduce the key notions of entropy, relative entropy (also called Kullback information in statistics) and mutual information. These quantities are concerned with the laws of random variables. Then, given a source, i.e. a random process $(X_n)_{n\geqslant 1}$, we introduce the notion of entropy rate, which will be the target quantity for the asymptotic compression rate that we will meet thereafter.

Definition 1.3 If P is a probability defined over a finite or countable set A, its *Shannon entropy* (or simply its entropy) is

$$H(P) = -\sum_{\omega \in A} P(\omega) \log_2 P(\omega).$$

The entropy may be equal to $+\infty$.

If X is a random variable with distribution P, we have

$$H(P) = E\big[-\log_2 P(X) \big].$$

Thus, this quantity is always positive or equal to zero.

The entropy of P will be denoted indifferently by $H(X)$ or $H(P)$; we will sometimes write "entropy of the random variable X" to refer to the entropy of its distribution P.

Definition 1.4 If P and Q are two probabilities over the same finite or countable set A, the *relative entropy* (or *Kullback information*) of P with respect to Q is

$$D(P \mid Q) = \sum_{\omega \in A} P(\omega) \log_2 \frac{P(\omega)}{Q(\omega)}.$$

The relative entropy may be equal to $+\infty$. This happens, for instance, when P is not absolutely continuous with respect to Q, i.e. when there exists an $\omega \in A$ such that $P(\omega) > 0$ and $Q(\omega) = 0$. When P is absolutely continuous with respect to Q, one may have $D(P \mid Q) = +\infty$ only if A is infinite.

If X is a random variable with values in A and distribution P, we have

$$D(P \mid Q) = E\Big[-\log_2 \frac{Q(X)}{P(X)} \Big].$$

The function $u \mapsto -\log_2 u$ being strictly convex, by Jensen's Inequality, $D(P \mid Q) \geqslant 0$, and $D(P \mid Q) = 0$ if and only if $P = Q$.

Definition 1.5 If (X, Y) is a random variable with values in a countable set and distribution P_{XY}, with marginals P_X and P_Y, the *mutual information* of X and Y is the relative entropy (or *Kullback information*) of their joint distribution with respect to the product of their marginal distributions:

$$I(X; Y) = D\big(P_{XY} \mid P_X \otimes P_Y\big) = E\Big(\log_2 \frac{P_{XY}(X, Y)}{P_X(X) P_Y(Y)} \Big).$$

Definition 1.6 If (X, Y) is a random variable with values in a countable set and if $P_{Y|X}$ is the conditional distribution of Y given X, the **conditional entropy** of Y given X is the expectation (in X) of the entropy of the conditional distribution $P_{Y|X}$:

$$H(Y|X) = E\left[H\left(P_{Y|X}\right)\right].$$

The entropy of a pair is computed by adding to the entropy of one of the random variables the conditional entropy of the other one, and this calculation is extended by induction to n-tuples of variables. The mutual information corresponds to what remains of the entropy when the conditional entropy is deducted.

In particular, the conditional entropy of Y given X is smaller than the entropy of Y; it is equal to the entropy of Y if and only if X and Y are independent random variables. The more random variables the are involved and the less dependent they are, the larger the entropy. It is maximal for the uniform distribution. This is expressed in the following properties:

1. $H(X, Y) = H(X) + H(Y|X)$,
2. $H(X_{1:n}) = H(X_1) + \sum_{i=2}^{n} H(X_i | X_{1:i-1})$,
3. $I(X; Y) = H(Y) - H(Y|X) = H(X) - H(X|Y) = H(X) + H(Y) - H(X, Y)$ and $I(X; Y) = 0$ if and only if X and Y are independent,
4. $H(X|Y) \leqslant H(X)$,
5. $H(X_{1:n}) \leqslant \sum_{i=1}^{n} H(X_i)$,
6. if X takes values in a finite A, then $H(X) \leqslant \log_2 |A|$.

Property 1 stems from the expression of the joint distribution of (X, Y) as the product of the marginal distribution of X and the conditional distribution of Y given X; Property 2 is obtained by induction. Property 3 stems from

$$\log_2 \frac{P_{XY}(X, Y)}{P_X(X) P_Y(Y)} = -\log_2 P_Y(Y) + \log_2 \frac{P_{XY}(X, Y)}{P_X(X)} = -\log_2 P_Y(Y) + \log_2 P_{Y|X}(Y).$$

Property 4 follows since the mutual information is always positive or zero. Combining Properties 2 and 4 gives Property 5. Finally, for Property 6, if A has m elements, by convexity of the function $u \mapsto u \log_2 u$:

$$\frac{1}{m} \sum_{\omega \in A} P(\omega) \log_2 P(\omega) \geqslant \frac{1}{m} \log_2 \frac{1}{m}.$$

Let (X, Y, Z) a random variable. Denote by $I(X; Y|Z)$ the expectation (in Z) of the mutual information of X and Y given Z. We have

$$I(X; (Y, Z)) = I(X; Z) + I(X; Y|Z) = I(X; Y) + I(X; Z|Y).$$

Indeed, by Property 3,

$$I(X; Z) + I(X; Y|Z) = H(X) - H(X|Z) + H(X|Z) - H(X|Y, Z).$$

The scheme $X \to Y \to Z$ is called *Markovian* if the distribution of Z given (X, Y) is equal to the distribution of Z given Y. Since

$$\frac{P_{(X,Z)|Y}}{P_{X|Y}P_{Z|Y}} = \frac{P_{Z|(X,Y)}}{P_{Z|Y}},$$

the variables X and Z are independent given Y if and only if the scheme $X \to Y \to Z$ is Markovian. In other words, we have $I(X; Z \mid Y) = 0$ if and only if $P_{Z|(Y,X)} = P_{Z|Y}$. It also follows that:

Proposition 1.3 *If the scheme $X \to Y \to Z$ is Markovian, then $I(X; Z) \leqslant I(X; Y)$.*

Definition 1.7 If $(X_n)_{n \in \mathbb{N}}$ is a process with distribution $\mathbb{P}$, the *entropy rate* of $\mathbb{P}$, if the limit exists, is given by,

$$H_*(\mathbb{P}) = \lim_{n \to +\infty} \frac{1}{n} H(X_{1:n})$$

and we will sometimes call it the *entropy rate* of the process $(X_n)_{n \in \mathbb{N}}$.

Let us recall that a process $(X_n)_{n \in \mathbb{N}}$, or $(X_n)_{n \in \mathbb{Z}}$, is *stationary* if, for all integers $m \geqslant 0$, and all integers $p \leqslant q$ (non-negative or possibly not, according to whether the sequence is indexed by $\mathbb{N}$ or by $\mathbb{Z}$), $(X_p, \ldots, X_q)$ has the same distribution as $(X_{p+m}, \ldots, X_{q+m})$.

Proposition 1.4 *If $(X_n)_{n \in \mathbb{N}}$ is a stationary process, then its entropy rate exists and*

$$H_*(\mathbb{P}) = \lim_{n \to +\infty} \frac{1}{n} H(X_{1:n}) = \inf_{n \in \mathbb{N}} \frac{1}{n} H(X_{1:n}) = \lim_{n \to +\infty} H(X_n \mid X_{1:n-1}).$$

Moreover, $H_*(\mathbb{P}) = +\infty$ if and only if $H(X_1) = +\infty$.

Proof (Proof of Proposition 1.4) For all integers n and m, by stationarity

$$H(X_{1:n+m}) = H(X_{n+1:n+m} \mid X_{1:n}) + H(X_{1:n})$$
$$\leqslant H(X_{n+1:n+m}) + H(X_{1:n}) = H(X_{1:m}) + H(X_{1:n}).$$

Letting $u_n = H(X_{1:n})$ for all integers n, we have shown that the non-negative sequence $(u_n)_{n \in \mathbb{N}}$ is sub-additive. By the classical lemma stated below, the sequence $(u_n)_{n \in \mathbb{N}}$ converges to its infimum. Thus $H_*(\mathbb{P})$ exists and is equal to

$$H_*(\mathbb{P}) = \lim_{n \to +\infty} \frac{1}{n} H(X_{1:n}) = \inf_{n \in \mathbb{N}} \frac{1}{n} H(X_{1:n}).$$

Hence, $H_*(\mathbb{P}) < +\infty$ if and only if there exists an n such that $H(X_{1:n}) < +\infty$. But $H(X_1) \leqslant H(X_{1:n}) \leqslant H(X_1) + \cdots + H(X_n) \leqslant nH(X_1)$ by stationarity, so that there exists an n such that $H(X_{1:n}) < +\infty$ if and only if $H(X_1) < +\infty$.

Also, $H(X_n \mid X_{1:n-1})$ is a decreasing sequence bounded below by 0, it thus has a limit. By Cesàro's Lemma, this limit is the same as that of

$$\frac{1}{n}H(X_1) + \frac{1}{n}\sum_{i=2}^{n} H(X_i \mid X_{1:i-1}) = \frac{1}{n}H(X_{1:n}). \qquad \square$$

Example 1.8

- If $\mathbb{P}$ is the distribution of the sequence $(X_n)_{n\in\mathbb{N}}$, where $(X_n)_{n\in\mathbb{N}}$ are independent random variables identically distributed according to distribution P, we have $H_*(\mathbb{P}) = H(P)$. Indeed, $H(X_{1:n}) = \sum_{i=1}^{n} H(X_i) = nH(P)$ for all integers n.
- If $\mathbb{P}$ is the distribution of a stationary Markov chain $(X_n)_{n\in\mathbb{N}}$, with stationary distribution μ and transition Π over a finite or countable set A,

$$H_*(\mathbb{P}) = -\sum_{(i,j)\in A^2} \mu_i \Pi_{i,j} \log_2 \Pi_{i,j}.$$

Indeed, for all integers n,

$$H(X_n \mid X_{1:n-1}) = H(X_n \mid X_{n-1})$$
$$= \sum_{i\in A} \mathbb{P}(X_{n-1} = i)\left[-\sum_{j\in A} \mathbb{P}(X_n = j \mid X_{n-1} = i) \log_2 \mathbb{P}(X_n = j \mid X_{n-1} = i)\right]$$
$$= -\sum_{(i,j)\in A^2} \mu_i \Pi_{i,j} \log_2 \Pi_{i,j}.$$

Lemma 1.5 *Let $(u_n)_{n\in\mathbb{N}}$ be a non-negative real sequence such that $u_{m+n} \leqslant u_m + u_n$, for all integers m and n. Then the sequence $(\frac{1}{n}u_n)_{n\geqslant 1}$ converges to its infimum.*

Proof (Proof of Lemma 1.5) Let $\ell = \inf_{n\geqslant 1} \frac{1}{n}u_n$. For all $\varepsilon > 0$, there exists a $p \geqslant 1$ such that $\ell \leqslant \frac{1}{p}u_p \leqslant \ell + \varepsilon$. For all integers n, by Euclidean division, there exist integers k and r such that $0 \leqslant r \leqslant p - 1$ and $n = kp + r$. We have $u_n \leqslant u_{kp} + u_r \leqslant ku_p + u_r$ by induction on k, thus for n large enough

$$\ell \leqslant \frac{u_n}{n} \leqslant \frac{k}{kp+r}u_p + \frac{u_r}{n} \leqslant \ell + \varepsilon + \frac{\max\{u_0, \ldots, u_{p-1}\}}{n} \leqslant \ell + 2\varepsilon. \qquad \square$$

1.3   Shannon Entropy and Compression

We will now see that the Shannon entropy of the random variable we are encoding is a lower bound for the expected code length. There is thus a limitation to compression, and this limitation is function of the distribution of the variable being encoded.

Theorem 1.6 *If f is uniquely decodable, if P is a probability over A, and if X has distribution P, then*

$$E\big(\ell[f(X)]\big) \geqslant H(P).$$

Proof (Proof of Theorem 1.6) Let us first assume that $H(P)$ is finite.

As indicated at the end of Sect. 1.1, we identify the length of a codeword with minus the logarithm of the probability of the encoded word by adding an element e to the set of words being encoded. Let then Q be the probability over $A \cup \{e\}$ such that $\ell(f(X)) = -\log_2 Q(X)$. Then

$$E\big(\ell[f(X)]\big) = E\big(-\log_2 Q(X)\big) = H(P) + D(P \mid Q) \geqslant H(P).$$

If $H(P)$ is infinite, then A is itself infinite. Let $(B_n)_{n \geqslant 1}$ be a (weakly) increasing sequence (inclusion-wise) of finite sets whose union is A. Applying the above inequality to the conditional distribution of X given the event $(X \in B_n)$, we obtain

$$\frac{E(\ell[f(X)]1_{X \in B_n})}{P(B_n)} \geqslant -\frac{1}{P(B_n)} \sum_{\omega \in B_n} P(\omega) \log_2 P(\omega) + \log_2 P(B_n)$$

and let n tend to infinity.

1.3.1 Shannon's Coding

If P is a probability over a finite or countable set A, for all $\omega \in A$, let

$$\lambda(\omega) := \big\lceil -\log_2 P(\omega) \big\rceil,$$

the smallest integer greater than or equal to $-\log_2 P(\omega)$. Then

$$2^{-\lambda(\omega)} \leqslant P(\omega) < 2^{-\lambda(\omega)+1}.$$

Now, thanks to Theorem 1.2, there exists a prefix code f whose length is given by λ. Since $\ell[f(\omega)] < -\log_2 P(\omega) + 1$ for all $\omega \in A$, we get:

Theorem 1.7 *If P is a probability over a finite or countable set and if X has distribution P, then there exists a prefix code such that*

$$E\big(\ell[f(X)]\big) < H(P) + 1.$$

Remarks 1.9

- The proof of Theorem 1.2 gives a method to construct a prefix code by binary expansion, but this code requires us to know P.
- If $(X_n)_{n\in\mathbb{N}}$ is a process over $\mathcal{X}^{\mathbb{N}}$, if for all integers n, P_n is the distribution of $X_{1:n}$, and if L_n^* denotes the infimum, over the set of prefix codes over $\mathcal{X}^n$, of the expected codeword length of $X_{1:n}$, then $\frac{1}{n}L_n^*$ is the optimal compression rate and it satisfies
$$\frac{H(P_n)}{n} \leqslant \frac{L_n^*}{n} \leqslant \frac{H(P_n)}{n} + \frac{1}{n}.$$

Thus if $\frac{1}{n}H(P_n)$ has a limit when n tends to infinity, this limit is the asymptotic optimal compression rate. If $(X_n)_{n\in\mathbb{N}}$ is a stationary process, its entropy rate is the asymptotic optimal compression rate.

1.3.2 Huffman's Coding

Huffman [1] discovered a simple algorithm to construct a code with optimal compression rate when the set to be encoded is finite.

Let A be a finite set of size m, denoted $A = \{\omega_1, \ldots, \omega_m\}$, and let $P = (p(\omega_i))_{i=1}^m$ be a probability over A. We assume that $\inf_i p(\omega_i) > 0$, because there is no point in encoding elements with zero probability (they can be withdrawn from A).

If f is a given prefix code on A and if $\ell_i = \ell[f(\omega_i)]$, for $i = 1, \ldots, m$, then the expected code length is $\sum_{i=1}^m \ell_i p(\omega_i)$. Since there exists an integer N such that $N \inf_i p(\omega_i) > \sum_{i=1}^m \ell_i p(\omega_i)$, in order to minimize the expected code length, it is sufficient to consider prefix codes g with lengths satisfying $\sup_i \ell[g(\omega_i)] \leqslant N$, which constitutes a finite set of codes.

Therefore, there exists a prefix code with optimal compression, i.e. a prefix code which minimizes $E[\ell(f(X))]$ over the set of prefix codes f. This prefix code with optimal compression is not necessarily unique. Huffman [1] discovered a simple algorithm to devise such a code, which we now describe.

To do so, we use the correspondence between prefix codes and binary trees. We construct a binary tree in which each leaf is associated to an element of A and corresponds to its codeword. This construction is made by induction. We start by ordering the elements of A according to their probability, say $p(\omega_1) \geqslant \cdots \geqslant p(\omega_m)$. Then, we take the last two elements in that order (those with smallest probability), and we assign each of them to a leaf, with an intermediary node as parent. We give weight $p(\omega_m) + p(\omega_{m-1})$ to that node (see the figure below).

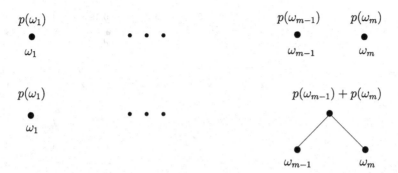

We re-order the weights and repeat. When only one node remains, the algorithm stops and we have a binary tree.

This algorithm constructs, in a finite number of steps, a prefix code such that

1. The two words with smallest probability have a code of equal length, differing in the last bit only.
2. Lengths are in the opposite order of probabilities.

This algorithm does not necessarily yield a unique code: when two weights are equal, we may have to choose the elements that we take to have the same parent node. For instance, with $A = \{a, b, c, d, e, f\}$ and the probability

$$
\begin{array}{ccccccc}
A & a & b & c & d & e & f \\
p & 0.25 & 0.20 & 0.20 & 0.15 & 0.15 & 0.05
\end{array}
$$

here are two possible Huffman's codes (each line in the graphical representations corresponds to a step of the grouping/re-ordering procedure):

- With code 1 (see Fig. 1.2), we obtain

$$g\,(a) = 01,\ g\,(b) = 10,\ g\,(c) = 11,\ g\,(d) = 001,\ g\,(e) = 0000,\ g\,(f) = 0001,$$

and the expected code length is

$$4 \times (0.05 + 0.15) + 3 \times 0.15 + 2 \times (0.2 + 0.2 + 0.25) = 2.55.$$

- With code 2 (see Fig. 1.3), we obtain:

$$g\,(a) = 01,\ g\,(b) = 10,\ g\,(c) = 000,\ g\,(d) = 001,\ g\,(e) = 110,\ g\,(f) = 111,$$

and the expected code length is

$$3 \times (0.05 + 0.15 + 0.15 + 0.2) + 2 \times (0.2 + 0.25) = 2.55.$$

Fig. 1.2 Code 1

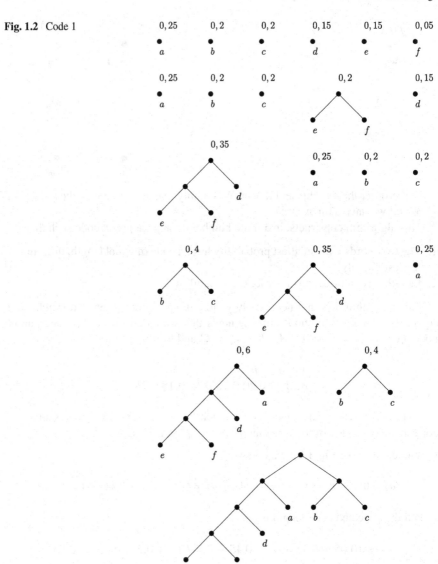

Theorem 1.8 *The code obtained by Huffman's algorithm is of optimal compression.*

Proof (Proof of Theorem 1.8) We start by showing that there exists a prefix code of optimal compression satisfying the indicated Properties 1 and 2 for the code produced by Huffman's algorithm. Let A be a set of size m whose elements are indexed in decreasing order of their probability ($p(\omega_1) \geqslant p(\omega_2) \geqslant \cdots$). First, a prefix code of optimal compression necessarily satisfies Property 2. Let f be a prefix code of optimal compression over A. Either f satisfies Property 1, or not. If f does not

Fig. 1.3 Code 2

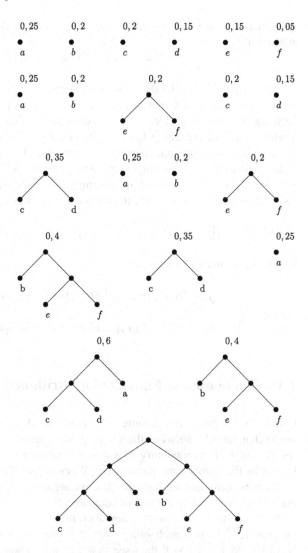

satisfy 1, either there exist $\omega_i \neq \omega_m$ such that $f(\omega_i)$ and $f(\omega_m)$ only differ in the last bit, and optimality implies that $p(\omega_i) = p(\omega_{m-1})$, or there is no $\omega_i \neq \omega_m$ such that $f(\omega_i)$ and $f(\omega_m)$ only differ in the last bit, in which case we can remove the last bit of $f(\omega_m)$ and still have a prefix code, which contradicts the optimality.

Let us now show by induction that the code produced by Huffman's algorithm is of optimal compression.

If $m = 2$, the result is trivial.

Assume that the theorem is true for all sets of cardinality m. Let A be a set of cardinality $m + 1$, P a distribution over A, the elements of A being indexed in decreasing order of their probability, and f be a code produced by Huffman's algorithm. Let

$$B = \{\omega_0\} \cup (A \setminus \{\omega_m, \omega_{m+1}\}),$$

and $Q = (q(\omega_i))_{1 \leq i \leq m}$ be the probability defined over B which coincides with P on $A \setminus \{\omega_m, \omega_{m+1}\}$, such that $q(\omega_0) = p(\omega_m) + p(\omega_{m+1})$. Let g be the code over B such that $g(\omega_i) = f(\omega_i)$ for $i < m - 1$ and $g(\omega_0)$ is $f(\omega_m)$ without the last symbol. Then g is a code stemming from Huffman's algorithm over B. By the induction hypothesis, g is of optimal compression over B. Let now f' be a code of optimal compression over A satisfying Properties 1 and 2. Define the code g' over B by $g'(\omega_i) = f'(\omega_i)$ for $i < m - 1$ and setting $g'(\omega_0)$ to be equal to $f'(\omega_m)$ without the last symbol as f' satisfies 2. By the induction hypothesis, we have

$$E_Q(\ell[g(\omega)]) \leqslant E_Q(\ell[g'(\omega)]) = E_P(\ell[f'(\omega)]) - (p(\omega_m) + p(\omega_{m+1})),$$

where ω is a random variable. But

$$E_Q(\ell[f(\omega)]) = E_P(\ell[f(\omega)]) - (p(\omega_m) + p(\omega_{m+1})).$$

Thus $E_P(\ell[f(\omega)]) \leqslant E_P(\ell[f'(\omega)])$ and f is a code of optimal compression over A.

1.4   Shannon-Fano-Elias Coding. Arithmetic Coding

Given a probability P over a finite or countable set A, we will describe a prefix code on the elements of A such that the length of the codeword for x is $1 + \lceil -\log_2 P(x) \rceil$, for all x in A. The probability P is then a parameter for the code, and needs to be known by the encoder and the decoder. When we say that we encode according to P, we now mean that we have an effective algorithm of implementation, provided the set we are encoding is finite or countable.

If $\mathcal{X}$ is a finite or countable alphabet, the set $\mathcal{X}^n$ is finite or countable for all integers n. If P_n is a probability over $\mathcal{X}^n$, we can, with this algorithm, encode according to P_n. We will then see that we can sequentially encode a word $x_{1:n}$. This way, if $\mathbb{P}$ is a probability over $\mathcal{X}^{\mathbb{N}}$, given by the sequence of its conditional distributions, we will be able to encode in a sequential way the words of $\mathcal{X}^*$: this is arithmetic coding.

Let us emphasize that the distribution according to which arithmetic coding is implemented is generally not the source distribution. The algorithmic aspect of coding and the modeling of the source distribution are thus handled separately.

We first describe the essence of arithmetic coding; and then explain its sequential implementation.

Let P be a probability over a finite or countable set A. The elements of A are labeled in an arbitrary way $\omega_1, \ldots, \omega_n, \ldots$ and ω_i is identified with the integer i. This allows us to define a *cumulative distribution function* F by

$$F(x) = P\,(U \leqslant x),$$

where U is the canonical variable over A. We then define

$$\widetilde{F}(x) = F(x) - \tfrac{1}{2}P(x) = P\,(U < x) + \tfrac{1}{2}P(x).$$

For all x in A, the codeword $f(x)$ is equal to the first $1 + \lceil -\log_2 P(x) \rceil$ terms in the binary expansion of $\widetilde{F}(x)$.

Let us now show that this is a prefix code. If $f(x)$ is a prefix of $f(y)$, then $P(x) \geqslant P(y)$ and the first $1 + \lceil -\log_2 P(x) \rceil$ terms in the binary expansion of $\widetilde{F}(x)$ and $\widetilde{F}(y)$ are the same. So

$$\left| \widetilde{F}(x) - \widetilde{F}(y) \right| \leqslant \tfrac{1}{2}P(x).$$

Now, since $P(y) > 0$,

- if $x < y$, then $\widetilde{F}(y) - \widetilde{F}(x) = P(x \leqslant X < y) + \tfrac{1}{2}P(y) - \tfrac{1}{2}P(x) > \tfrac{1}{2}P(x)$,
- if $x > y$, then $\widetilde{F}(x) - \widetilde{F}(y) = P(y \leqslant X < x) + \tfrac{1}{2}P(x) - \tfrac{1}{2}P(y) > \tfrac{1}{2}P(x)$.

Therefore $f(x)$ cannot be a prefix of $f(y)$ for $x \neq y$.

Moreover, if X is a random variable with distribution P over A, then

$$E\big(\ell[f(X)]\big) < H(P) + 2.$$

In other words, if this coding procedure—called *Shannon-Fano-Elias coding*—is implemented according to the distribution of X, it achieves the optimal compression, up to at most an additive factor of 2.

Let us now describe how to encode in a sequential way words of $\mathscr{X}^*$ according to a probability distribution given by the sequence of its conditional distributions, by applying Shannon-Fano-Elias coding over $A = \mathscr{X}^n$ for all integers n.

This sequential coding is called *arithmetic coding*. Here we assume that the alphabet $\mathscr{X}$ is finite. The order on $\mathscr{X}$ induces the lexicographic order on $\mathscr{X}^n$: $x_{1:n} > y_{1:n}$ means that $x_i > y_i$ for the first i such that $x_i \neq y_i$.

Let $\mathbb{P}$ be a probability over $\mathscr{X}^{\mathbb{N}}$ and $(U_n)_{n \geqslant 1}$ the canonical process. For all integers n, denote by $\mathbb{P}_n$ the distribution of $U_{1:n}$ under $\mathbb{P}$ and by F_n its cumulative distribution function. We then encode $x_{1:n}$ by the first $1 + \lceil -\log_2 \mathbb{P}_n(x_{1:n}) \rceil$ terms in the binary expansion of $F_n(x_{1:n}) - \tfrac{1}{2}\mathbb{P}_n(x_{1:n})$.

One may compute $i(x_{1:n}) = \mathbb{P}_n(U_{1:n} < x_{1:n})$ and $s(x_{1:n}) = \mathbb{P}_n(U_{1:n} \leqslant x_{1:n})$ in a recursive way by

$$i(x_{1:n+1}) = i(x_{1:n}) + \mathbb{P}_n(x_{1:n})\mathbb{P}\big(U_{n+1} < x_{n+1} \,|\, U_{1:n} = x_{1:n}\big),$$

$$s(x_{1:n+1}) = i(x_{1:n}) + \mathbb{P}_n(x_{1:n})\mathbb{P}\big(U_{n+1} \leqslant x_{n+1} \mid U_{1:n} = x_{1:n}\big).$$

Indeed

$$(U_{1:n+1} < x_{1:n+1}) = (U_{1:n} < x_{1:n}) \cup \big(\{U_{1:n} = x_{1:n}\} \cap \{U_{n+1} < x_{n+1}\}\big),$$

$$(U_{1:n+1} \leqslant x_{1:n+1}) = (U_{1:n} < x_{1:n}) \cup \big(\{U_{1:n} = x_{1:n}\} \cap \{U_{n+1} \leqslant x_{n+1}\}\big).$$

The interval $I_n = [i(x_{1:n}), s(x_{1:n})]$ being of length $\mathbb{P}_n(x_{1:n})$, its midpoint is $F_n(x_{1:n}) - \frac{1}{2}\mathbb{P}_n(x_{1:n})$. Consequently, the sequence of the first $1 + \lceil -\log_2 \mathbb{P}_n(x_{1:n})\rceil$ terms in its binary expansion forms the codeword of $x_{1:n}$.

Example 1.10 For $\mathscr{X} = \{1, 2, 3\}$, $n = 2$ and the coding distribution $P_2 = P_1 \otimes P_1$, where $P_1(1) = 0.3$, $P_1(2) = 0.6$ and $P_1(3) = 0.1$, the decomposition of the interval $[0, 1]$ is as follows:

With the preceding notation: if $x_{1:2} = 22$, $I_2 = [0.48, 0.84]$, the midpoint is 0.66. Since $P_2(22) = 0.36$ and $\lceil -\log_2 0.36\rceil + 1 = 3$, the codeword is 101. To see why, we successively cut the interval $[0, 1]$ in half three times, and examine whether 0.66 lies to the left or to the right of the successive half-cuts:

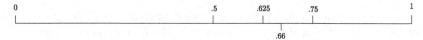

Implementing this operation for each of the possible values of $x_{1:2}$, we arrive at the code whose tree is depicted in Fig. 1.4.

The tree is far from being complete; this example shows that each codeword may well be too long by one or two bits. If the distribution of $X_{1:2}$ is P_2, then the expected code length is ≈ 4.2 whereas the entropy is ≈ 2.6.

The fact that one can code a sequence of letters by sequentially computing the intervals I_n is quite remarkable. Let us notice that those intervals are decreasing. Let us also observe that for $k < n$:

- If $I_k \subset [0, \frac{1}{2}]$, then $I_j \subset [0, \frac{1}{2}]$ for all $j > k$ and we know that the first term of the codeword of $x_{1:n}$ will be 0.
- If $I_k \subset [\frac{1}{2}, 1]$, then $I_j \subset [\frac{1}{2}, 1]$ for all $j > k$ and we know that the first term of the codeword of $x_{1:n}$ will be 1.
- If on the other hand, $I_k \subset [\frac{1}{4}, \frac{3}{4}]$, we don't know the first term, but we know that the first and second term will be different.

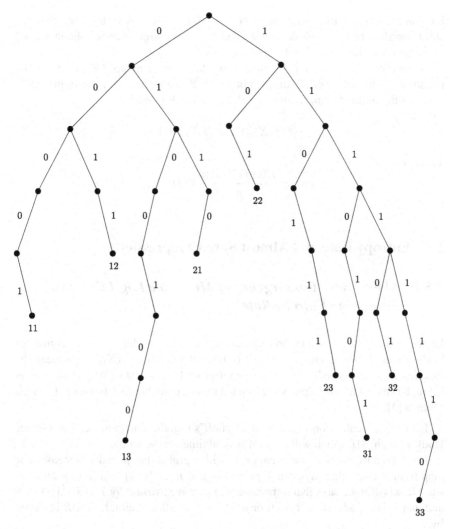

Fig. 1.4 Tree associated to Example 1.10

By iteratively applying these three properties: replacing I_k by $2I_k$ in the first case, by $2I_k - 1$ in the second case, and by $2I_k - \frac{1}{2}$ in the third case, we sequentially determine the terms of the codeword of $x_{1:n}$, by reducing (possibly iterating several times) to an interval of width larger than $\frac{1}{4}$. We finally end the codeword when we reach $k = n$.

Similarly, there exists a very simple and iterative way of decoding the codeword $y_{1:m}$ of a word $x_{1:n}$ of length n. Let $\tau \in [0, 1]$ be equal to $\sum_{j=1}^{m} y_j 2^{-j}$. We start by cutting the interval $[0, 1]$ according to the distribution of P_1 and observe to which of those sub-intervals τ belongs: the associated symbol x_1 is the first of the decoded message $x_{1:n}$. We then compute $I_1 = [i(x_1), s(x_1)]$, as the encoder, and cut this interval according to the distribution $\mathbb{P}(U_2 = \cdot \,|\, U_1 = x_1)$. Observing to which of the sub-intervals τ belongs, we deduce x_2, and so on. We stop when we obtain a word of length n (the integer n thus has to be known).

To conclude this section, if we encode the stationary sequence $(X_n)_{n \in \mathbb{N}}$ with distribution $\mathbb{P}$ using arithmetic coding according to $\mathbb{P}$, arithmetic coding asymptotically achieves the optimal compression rate. Indeed, for all n, we have

$$E\big(\ell[f(X_{1:n})]\big) < H(\mathbb{P}_n) + 2,$$

and thus

$$\limsup_{n \to +\infty} \frac{E(\ell[f(X_{1:n})])}{n} \leqslant H_*(\mathbb{P}).$$

1.5 Entropy Rate and Almost Sure Compression

1.5.1 Almost Sure Convergence of Minus the Log-Likelihood Rate to the Entropy Rate

Under weak conditions (stationarity and ergodicity), $-\frac{1}{n} \log_2 \mathbb{P}(X_{1:n})$ converges $\mathbb{P}$-almost surely to the entropy rate $H_*(\mathbb{P})$. By the notation $\mathbb{P}(X_{1:n})$, we mean the function $\mathbb{P}_n(\cdot)$ evaluated at $X_{1:n}$. The proof given here is that of Algoet and Cover [2] in the case of a finite alphabet, extended to the countable case thanks to Chung's Lemma [3].

Let us first recall a consequence of Birkhoff's Ergodic Theorem (see for instance Dudley [4, ch. 8]), which will be used several times in the sequel.

Let Z be a measurable space endowed with a probability μ and T a measurable map from Z to Z. One says that T preserves μ if $\mu(T^{-1}(B)) = \mu(B)$ for all measurable sets B. One says that a measurable set B is invariant by T if $T^{-1}(B) = B$ and that T is ergodic if $\mu(B) = 0$ or $\mu(B) = 1$ for all measurable sets B invariant by T.

Applying this to $Z = \mathscr{X}^{\mathbb{N}}$, $\mu = \mathbb{P}$ and T given by $T[(x_n)_{n \geqslant 1}] = (x_{n+1})_{n \geqslant 1}$, one says that the process $(X_n)_{n \geqslant 1}$ with distribution $\mathbb{P}$ is stationary ergodic if T preserves μ and is ergodic.

A remarkable result, which is a consequence of Birkhoff's Ergodic Theorem, is that the mean over time and the mean over space almost surely coincide asymptotically, in the following sense.

Theorem 1.9 *If $(X_n)_{n \in \mathbb{N}}$ is an ergodic and stationary process of distribution $\mathbb{P}$ with values in $\mathscr{X}$ and if h is a map from $\mathscr{X}^{\mathbb{N}}$ to $\mathbb{R}$ such that $|h((X_n)_{n \in \mathbb{N}})|$ has finite expectation, then $\frac{1}{n} \sum_{i=1}^{n} h\left((X_{n+i})_{n \in \mathbb{N}}\right)$ converges $\mathbb{P}$-a.s. and in $L_1(\mathbb{P})$ to $E[h((X_n)_{n \in \mathbb{N}})]$ as n tends to infinity.*

If $(X_n)_{n\in\mathbb{N}}$ is a sequence of independent and identically distributed random variables, then $(X_n)_{n\in\mathbb{N}}$ is an ergodic and stationary process. If $(X_n)_{n\in\mathbb{N}}$ is a stationary, irreducible and positive recurrent Markov chain over a finite or countable state space, then $(X_n)_{n\in\mathbb{N}}$ is an ergodic and stationary process.

We can now state the following theorem :

Theorem 1.10 (Shannon-Breiman-McMillan) *If $(X_n)_{n\in\mathbb{N}}$ is a stationary and ergodic process with distribution $\mathbb{P}$, values in a finite or countable set $\mathscr{X}$ and finite entropy rate $H_*(\mathbb{P})$ (or equivalently, such that $H(X_1) < +\infty$), then $-\frac{1}{n}\log_2 \mathbb{P}(X_{1:n})$ converges $\mathbb{P}$-a.s. to $H_*(\mathbb{P})$ as n tends to infinity.*

Proof (Proof of Shannon-Breiman-McMillan Theorem) The proof is decomposed into three steps. We first extend, by stationarity, the distribution $\mathbb{P}$ to sequences indexed by $\mathbb{Z}$. We can then define a conditional distribution given an infinite past and identify the entropy rate with the entropy of the distribution of the present given the past. We then introduce the approximation of $\mathbb{P}$ by the distribution $\mathbb{P}^k$ of a Markov chain of order k, which allows us to use the result of ergodic theory stated in Theorem 1.9. Finally, we show that the infinite past is retrieved by letting k tend to infinity.

This idea (approximating $\mathbb{P}$ by the distribution $\mathbb{P}^k$ of a Markov chain of order k) will also be invoked in Chap. 2 to show that some codes are universal.

First step: extension by stationarity of the distribution $\mathbb{P}$ to sequences indexed by $\mathbb{Z}$. The distribution of $(X_n)_{n\in\mathbb{Z}}$ is defined by Kolmogorov's Extension Theorem and by the fact that the distribution of $X_{k:m}$ is equal to the distribution of $X_{1:m-k+1}$ for all $k, m \in \mathbb{Z}, k < m$.

For all x in $\mathscr{X}$, $\mathbb{P}(X_0 = x \mid X_{-n:-1})$ is a bounded non-negative martingale, and thus converges $\mathbb{P}$-a.s. and in $L_1(\mathbb{P})$ to what is denoted $\mathbb{P}(X_0 = x \mid X_{-\infty:-1})$. Since $\mathscr{X}$ is finite or countable, $-\log_2 \mathbb{P}(X_0 \mid X_{-n:-1})$ converges $\mathbb{P}$-a.s. to $-\log_2 \mathbb{P}(X_0 \mid X_{-\infty:-1})$ and in $L_1(\mathbb{P})$ if the sequence is equi-integrable, using the following lemma, whose proof is given at the end of the section.

Lemma 1.11 (Chung [3]) *If $\mathscr{X}$ is countable and if $H(X_0)$ is finite, then*

$$E\Big[\sup_{n\in\mathbb{N}^*} -\log_2 \mathbb{P}(X_0 \mid X_{-n:-1}) \Big] < +\infty.$$

We may thus define

$$H^\infty = E\big(-\log_2 \mathbb{P}(X_0 \mid X_{-\infty:-1})\big).$$

Defining

$$H^k = E\big(-\log_2 \mathbb{P}(X_{k+1} \mid X_{1:k})\big) = E\big(-\log_2 \mathbb{P}(X_0 \mid X_{-k:-1})\big),$$

H^k decreases to $H_*(\mathbb{P})$ as k tends to infinity and H^k also decreases to H^∞, which is therefore equal to $H_*(\mathbb{P})$.

Let us define $\log_2 \mathbb{P}(X_{1:n} \mid X_{-\infty:-1})$ and $\log_2 \mathbb{P}(X_n \mid X_{-\infty:n-1})$ in the same way $\mathbb{P}$-a.s. and in $L_1(\mathbb{P})$, for all integers n.

Second step: definition of the probability $\mathbb{P}^k$, approximation of $\mathbb{P}$ by the distribution of a Markov chain of order k. For all integers n and for all $x_{1:n} \in \mathcal{X}^n$, let

$$\mathbb{P}^k(x_{1:n}) = \begin{cases} \mathbb{P}(X_{1:n} = x_{1:n}) & \text{if } n \leqslant k, \\ \mathbb{P}(X_{1:k} = x_{1:k}) \prod_{i=k+1}^{n} \mathbb{P}(X_i = x_i \mid X_{i-k:i-1} = x_{i-k:i-1}) & \text{if } n > k. \end{cases}$$
(1.1)

Under $\mathbb{P}^k$, $(X_n)_{n \in \mathbb{N}}$ is a homogeneous Markov chain of order k, i.e. under $\mathbb{P}^k$, for all $n \geqslant k$, the conditional distribution of X_{n+1} given $X_{1:n}$ is equal to the conditional distribution of X_{n+1} given $X_{n-k+1:n}$.

For all $n > k$,

$$-\frac{1}{n} \log_2 \mathbb{P}^k(X_{1:n}) = -\frac{1}{n} \log_2 \mathbb{P}(X_{1:k}) - \frac{1}{n} \sum_{i=k+1}^{n} \log_2 \mathbb{P}(X_i \mid X_{i-k:i-1})$$

tends $\mathbb{P}$-a.s. to H^k by Theorem 1.9.

Similarly,

$$-\frac{1}{n} \log_2 \mathbb{P}(X_{1:n} \mid X_{-\infty:0}) = -\frac{1}{n} \sum_{i=1}^{n} \log_2 \mathbb{P}(X_i \mid X_{-\infty:i-1})$$

tends $\mathbb{P}$-a.s. to H^∞ by Theorem 1.9.

Third step (sandwich): By Markov's Inequality,

$$\mathbb{P}\left(\frac{1}{n} \log_2 \frac{\mathbb{P}^k(X_{1:n})}{\mathbb{P}(X_{1:n})} \geqslant 2\frac{\log_2 n}{n}\right) \leqslant \frac{1}{n^2}.$$

Borel-Cantelli Lemma then allows us to infer that, $\mathbb{P}$-a.s.

$$\limsup_{n \to +\infty} \frac{1}{n} \log_2 \frac{\mathbb{P}^k(X_{1:n})}{\mathbb{P}(X_{1:n})} \leqslant 0.$$

Thus, for all integers k, $\mathbb{P}$-a.s.

$$\limsup_{n \to +\infty} -\frac{1}{n} \log_2 \mathbb{P}(X_{1:n}) \leqslant H^k.$$

Similarly,

$$\mathbb{P}\left(\frac{\mathbb{P}(X_{1:n})}{\mathbb{P}(X_{1:n} \mid X_{-\infty:0})} \geqslant n^2\right) = E\left[\mathbb{P}\left(\frac{\mathbb{P}(X_{1:n})}{\mathbb{P}(X_{1:n} \mid X_{-\infty:0})} \geqslant n^2 \mid X_{-\infty:0}\right)\right] \leqslant \frac{1}{n^2},$$

thus, $\mathbb{P}$-a.s.

$$\limsup_{n \to +\infty} \frac{1}{n} \log_2 \frac{\mathbb{P}(X_{1:n})}{\mathbb{P}(X_{1:n} \mid X_{-\infty:0})} \leqslant 0,$$

and thus, $\mathbb{P}$-a.s.

$$\liminf_{n \to +\infty} -\frac{1}{n} \log_2 \mathbb{P}(X_{1:n}) \geqslant H^\infty.$$

The proof is concluded thanks to the first step letting k tend to infinity.

Proof (Proof of Chung's Lemma 1.11) For all integers $k \geqslant 1$, let

$$Z_k = -\log_2 \mathbb{P}(X_0 \mid X_{-k:-1}).$$

Define $\overline{Z} = \sup_{k \in \mathbb{N}^*} Z_k$. Then

$$E\left(\overline{Z}\right) \leqslant \sum_{m=0}^{\infty} \mathbb{P}(\overline{Z} \geqslant m) = \sum_{m=0}^{\infty} \sum_{k=1}^{\infty} \mathbb{P}\left(A_k(m)\right),$$

where

$$A_k(m) = \left\{ \sup_{j<k} Z_j < m, \ Z_k \geqslant m \right\}, \ k \geqslant 1, \ A_1(m) = \{Z_1 \geqslant m\}.$$

We order $\mathscr{X}$ in a decreasing way according to the distribution of X_0 and identify $\mathscr{X}$ with $\mathbb{N}^*$ in such a way that the sequence $(\mathbb{P}(X_0 = n))_{n \in \mathbb{N}^*}$ is decreasing. Let us then define for all integers i and for all $k \geqslant 1$

$$A_k^i(m) = \left\{ \sup_{j<k} -\log_2 \mathbb{P}(X_0 = i \mid X_{-j:-1}) < m, \ -\log_2 \mathbb{P}(X_0 = i \mid X_{-k:-1}) \geqslant m \right\}$$

and $A_1^i(m) = \left\{ -\log_2 \mathbb{P}(X_0 = i \mid X_{-1}) \geqslant m \right\}$. Then

$$\mathbb{P}\left(A_k^i(m) \cap (X_0 = i)\right) = E\left[1_{A_k^i(m)} E(1_{X_0=i} \mid X_{-k:-1})\right] \leqslant \frac{1}{2^m} \mathbb{P}\left(A_k^i(m)\right).$$

Let h be a function from $\mathbb{N}$ to $\mathbb{R}^+$. Then

$$\sum_{k=1}^{\infty} \sum_{i \leqslant h(m)} \mathbb{P}\left(A_k^i(m) \cap (X_0 = i)\right) \leqslant \frac{1}{2^m} \sum_{i \leqslant h(m)} \sum_{k=1}^{\infty} \mathbb{P}\left(A_k^i(m)\right) \leqslant \frac{h(m)}{2^m},$$

$$\sum_{k=1}^{\infty} \sum_{i > h(m)} \mathbb{P}\left(A_k^i(m) \cap (X_0 = i)\right) \leqslant \sum_{i > h(m)} \mathbb{P}(X_0 = i),$$

so that if $N(i)$ is the number of integers m such that $h(m) < i$:

$$\sum_{m=0}^{\infty}\sum_{k=1}^{\infty}\mathbb{P}\big(A_k(m)\big) \leqslant \sum_{m=0}^{\infty}\frac{h(m)}{2^m} + \sum_{i=1}^{\infty}N(i)\mathbb{P}(X_0 = i).$$

Taking $h(m) = 2^m/(m+1)^2$, there exist A and B such that $N(i) \leqslant A\log_2 i + B$, so that

$$\sum_{m=0}^{\infty}\sum_{k=1}^{\infty}\mathbb{P}\big(A_k(m)\big) \leqslant \sum_{m=0}^{\infty}\frac{1}{(m+1)^2} + A\sum_{i=1}^{\infty}(\log_2 i)\mathbb{P}(X_0 = i) + B.$$

Since the $\mathbb{P}(X_0 = i)$ form a decreasing sequence, we have $i\,\mathbb{P}(X_0 = i) \leqslant 1$ for all integers $i \geqslant 1$. Thus $\log_2 i + \log_2 \mathbb{P}(X_0 = i) \leqslant 0$ and we obtain

$$\sum_{i=1}^{\infty}(\log_2 i)\mathbb{P}(X_0 = i) \leqslant -\sum_{i=1}^{\infty}\mathbb{P}(X_0 = i)\log_2 \mathbb{P}(X_0 = i) = H(X_0).$$

We thus have

$$E\overline{(Z)} \leqslant \sum_{m=0}^{\infty}\frac{1}{(m+1)^2} + AH(X_0) + B < +\infty. \qquad \square$$

1.5.2   Almost Sure Compression Rate

We have seen that the entropy rate is a lower bound for the compression rate of uniquely decodable codes, measured as the limit in n (the length of the encoded word) of the expected code length normalized by n. We will strengthen this result by bounding from below, asymptotically almost surely, the code length normalized by n by the entropy rate. The proof will rely on the Kraft–McMillan Inequality and the result is thus established for uniquely decodable codes. We will see later that it can be extended to all lossless codes, with a simple transformation of a lossless code into a prefix code.

Theorem 1.12 *If f is uniquely decodable from $\mathscr{X}^*$ into $\{0,1\}^*$ and if $\mathbb{P}$ is the distribution of a stationary ergodic process over $\mathscr{X}^{\mathbb{N}}$ with finite entropy rate, then $\mathbb{P}$-a.s.*

$$\liminf_{n\to+\infty}\frac{\ell[f(X_{1:n})]}{n} \geqslant H_*(\mathbb{P}).$$

Proof (Proof of Theorem 1.12) For all sequence $(\alpha_n)_{n\in\mathbb{N}}$, we have, by Theorem 1.1:

$$\mathbb{P}\big(\ell[f(X_{1:n})] \leqslant -\log_2 \mathbb{P}(X_{1:n}) - \alpha_n\big) = \sum_{x_{1:n}:\ell[f(x_{1:n})]\leqslant -\log_2 \mathbb{P}(x_{1:n})-\alpha_n} \mathbb{P}(x_{1:n})$$

$$\leqslant \sum_{x_{1:n}} 2^{-\ell[f(x_{1:n})]} 2^{-\alpha_n} \leqslant 2^{-\alpha_n}.$$

Taking $\alpha_n = 2\log_2 n$ and appealing to Borel-Cantelli Lemma, we infer that $\mathbb{P}$-a.s. for n large enough,

$$\ell\big[f(X_{1:n})\big] \geqslant -\log_2 \mathbb{P}(X_{1:n}) - 2\log_2 n,$$

and thus, by Theorem 1.10, $\mathbb{P}$-a.s.

$$\liminf_{n\to +\infty} \frac{\ell[f(X_{1:n})]}{n} \geqslant \liminf_{n\to +\infty} \frac{-\log_2 \mathbb{P}(X_{1:n})}{n} = H_*(\mathbb{P}). \qquad \square$$

Let us now extend this result to a sequence of lossless codes, encoding, for all integers n, the words of $\mathscr{X}^n$. Let f_n be a lossless code over $\mathscr{X}^n$. In other words, $(f_n)_{n\geq 1}$ is a sequence of injections from $\mathscr{X}^n$ to $\{0,1\}^*$. Assume that $\mathscr{E}$ is a given prefix code from the set of non-negative integers into $\{0,1\}^*$. For all $x_{1:n}$ in $\mathscr{X}^*$, set:

$$f(x_{1:n}) = \mathscr{E}(n) \cdot \mathscr{E}\big(\ell[f_n(x_{1:n})]\big) \cdot f_n(x_{1:n}). \qquad (1.2)$$

The code f is then a prefix code: indeed, the first decoded integer indicates which code to use, and the second decoded integer indicates the length of the codeword to be decoded with the code indicated by the first decoded integer.

We will see that Elias (hence the notation $\mathscr{E}$) proposed a prefix code of the integers satisfying $l[\mathscr{E}(n)] = \log_2 n + o(\log_2 n)$. Using this code and appealing to Theorem 1.12, we have the following theorem.

Theorem 1.13 *If $(f_n)_{n\geqslant 1}$ is a sequence of lossless codes over $(\mathscr{X}^n)_{n\geqslant 1}$ and if $\mathbb{P}$ is the distribution of an ergodic stationary process over $\mathscr{X}^\infty$ with finite entropy rate $H_*(\mathbb{P})$, then $\mathbb{P}$-a.s.*

$$\liminf_{n\to +\infty} \frac{\ell[f_n(X_{1:n})]}{n} \geqslant H_*(\mathbb{P}).$$

Proof (Proof of Theorem 1.13) If $H_*(\mathbb{P}) = 0$, there is nothing to prove. Assume $H_*(\mathbb{P}) > 0$. Let f be the associated prefix code through (1.2). By Theorem 1.12,

$$\liminf_{n\to +\infty} \frac{\ell[f(X_{1:n})]}{n} \geqslant H_*(\mathbb{P}). \qquad (1.3)$$

But $\ell[f(X_{1:n})] = \ell[f_n(X_{1:n})] + \ell[\mathscr{E}(n)] + \ell[\mathscr{E}(\ell[f_n(X_{1:n})])]$. Since

$$\liminf_{n\to +\infty} \frac{\ell[\mathscr{E}(n)]}{n} = 0 \quad \text{and} \quad \ell\big[\mathscr{E}\big(\ell[f_n(X_{1:n})]\big)\big] = O\big(\log_2 \ell[f_n(X_{1:n})]\big),$$

we have that $\ell[f_n(X_{1:n})]$ tends $\mathbb{P}$-a.s. to $+\infty$ as n tends to $+\infty$. Therefore

$$\ell[\mathscr{E}(\ell[f_n(X_{1:n})])] = o(\ell[f_n(X_{1:n})]),$$

and the theorem easily follows from (1.3).

1.6   Notes

One may find a great part of the results presented in this chapter in the book by Cover and Thomas [5], together with a detailed bibliography. One has to mention Shannon, who is at the origin of a large part of these ideas and who proved the two fundamental theorems of communication: the Source Coding Theorem [6] and the Channel Coding Theorem (which will not be discussed in this book). Whereas source coding is concerned with data compression, channel coding aims at representing coded data in a way that is robust to transmission errors, in the sense that this representation should allow us to reconstruct the coded message transmitted with a controlled error rate.

Arithmetic coding was developed in the 1970s and 1980s. One may find its origin in (unpublished) works of Elias, and subsequently in Rissanen [7] and Pasco [8]. The section on arithmetic coding benefited from the clarity of the text of Aurélien Garivier [9].

The notions of entropy, relative entropy, mutual information and conditional entropy can be defined for probability distributions over more general spaces. In Sects. 2.2.3 and 2.3.1, we will use relative entropies for distributions over complete separable metric spaces. We will then give the definitions and general properties that shall be used.

Shannon [6] first established Theorem 1.10 for memoryless sources (sequences of i.i.d. random variables) and Markovian sources; he then formulated it for stationary ergodic sources. McMillan [10] proved L_1 convergence (and thus convergence in probability) of minus the normalized log-likelihood to the entropy rate for all stationary ergodic sources. Breiman [11] then obtained almost sure convergence for sources with values in finite alphabets. This theorem thus takes the name of the Shannon-Breiman-McMillan Theorem.

Generally speaking, if $\mathbb{P}$ is a distribution over $\mathscr{X}^{\mathbb{N}}$, for $\mathscr{X}$ a complete separable metric space, one may ask whether, when and to what, $-\frac{1}{n}\log_2\mathbb{P}(X_{1:n})$ converges, when the sequence $(X_n)_{n\in\mathbb{N}}$ has distribution $\mathbb{Q}$. When this sequence is ergodic and stationary, we call relative divergence rate (or relative entropy rate) of $\mathbb{Q}$ with respect to $\mathbb{P}$ the difference between this limit, when it exists, and $H_*(\mathbb{Q})$. Barron [12] showed a convergence result $\mathbb{Q}$-a.s., and in particular that this limit exists, under the assumption that $\mathbb{P}$ is the distribution of a stationary Markov chain. A counter-example from Kieffer [13, 14] shows that, without the Markovian assumption, this result might not be true and the relative entropy rate might not be defined.

References

1. D. Huffman, A method for the construction of minimum redundancy codes. Proc. IRE **40**, 1098–1101 (1952)
2. P. Algoet, T. Cover, A sandwich proof of the Shannon-McMillan-Breiman theorem. Annals of Prob. **16**, 899–909 (1988)
3. K. Chung, A note on the ergodic theorem of information theory. Annals of Math. Stat. **32**, 612–614 (1961)
4. R. Dudley, *Real analysis and probability*, 2nd edn. (Cambridge University Press, New York, 2002)
5. T.M. Cover, J.A. Thomas, *Elements of Information Theory*. Wiley series in telecommunications (Wiley, New York, 1991)
6. C. Shannon, A mathematical theory of communication. Bell Sys. Tech. J. **27**(379–423), 623–656 (1948)
7. J. Rissanen, Generelized Kraft inequality and arithmetic coding. IBM J. Res. Devl. **20**, 20–198 (1976)
8. R. Pasco. Source coding algorithms for fast data compression. Ph.D. Thesis, Stanford Univ (1976)
9. A. Garivier. Codage universel: la méthode arithmétique. Texte de préparation à l'agrégation (2006)
10. B. McMillan, The basic theorems of information theory. Ann. Math. Stat. **24**, 196–219 (1953)
11. L. Breiman, The individual ergodic theorem of information theory. Ann. Math. Stat. **28**, 809–811 (1957)
12. A. Barron, The strong ergodic theorem for densities: generalized Shannon-McMillan-Breiman theorem. Annals Probab. **13**, 1292–1303 (1985)
13. J. Kieffer, A counter-example to Perez's generalization of Shannon-McMillan theorem. Annals Probab. **1**, 362–364 (1973)
14. J. Kieffer, Correction to "a counter-example to Perez's generalization of Shannon-McMillan theorem". Annals of Probab. **4**, 153–154 (1976)

Chapter 2
Universal Coding on Finite Alphabets

Abstract Now that we have seen that the compression rate is lower-bounded by the entropy rate of the source, it is natural to wonder if there exist codes whose asymptotic compression rate is equal to the entropy rate of the source, whatever the source. This is possible when $\mathscr{X}$ is finite, provided the sources are ergodic and stationary. We will see in the next chapter (Chap. 3) that it is not the case if $\mathscr{X}$ is infinite. We will then be interested in the speed of convergence of the compression rate to the entropy rate of the source. This is where the connection with statistical methods based on maximum likelihood and Bayesian inference will appear. We will see that for parametric classes, the speed of convergence is lower-bounded by $\log_2 n$ times half the number of parameters.

Definition 2.1 A code f is *weakly universal* for the class $\mathscr{C}$ of laws over $\mathscr{X}^{\mathbb{N}}$ if for all laws $\mathbb{P}$ in $\mathscr{C}$, $\mathbb{P}$-a.s.,

$$\limsup_{n \to +\infty} \frac{\ell[f(X_{1:n})]}{n} \leqslant H_*(\mathbb{P}).$$

If $\mathscr{X}$ is finite, there exists a weakly universal lossless code over the class of ergodic and stationary processes. An example is Lempel-Ziv coding, which will be presented in the next section. In the remainder of the chapter, we will see a second example of a weakly universal lossless code over the class of ergodic and stationary processes, which is based on the approximation of stationary ergodic processes by Markov chains of order k and on a coding method related to Bayesian statistical methods. This code achieves the minimal speed for classes of memoryless sources, that is, formed by i.i.d. random variables, and for Markovian sources. Few things are known about non-parametric classes of sources. We will discuss the particular example of binary renewal sources, for which it is possible to precisely evaluate the minimal speed of convergence of the compression rate to the entropy rate of the source, and we will show that the Bayesian-type code which is optimal for Markovian sources is still optimal, up to a $\log_2 n$ factor for renewal sources.

© Springer International Publishing AG, part of Springer Nature 2018   29
É. Gassiat, *Universal Coding and Order Identification by Model Selection Methods*, Springer Monographs in Mathematics,
https://doi.org/10.1007/978-3-319-96262-7_2

2.1 Lempel-Ziv Coding

We describe a code proposed by Abraham Lempel and Jacob Ziv [1]. There actually exist several codes referred to as Lempel-Ziv codes, but they are all based on the crucial idea of using repetitions in the word in order to encode it in an efficient way. The compression software `gzip` uses Lempel-Ziv coding.

The idea is to decompose the word $x_{1:n}$ into sequences such that "the next sequence is the smallest new sequence":

$$x_{1:n} = w(1) \ldots w(2) \ldots w(c(x_{1:n})) \cdot v.$$

Here, $c(x_{1:n})$ is the number of distinct sequences in the decomposition. This decomposition is defined as follows:

- $w(1) = x_1$;
- if $w(1) \ldots w(j) = x_{1:n_j}$, either $x_{n_j+1} \notin \{w(1), \ldots, w(j)\}$ and then $w(j+1) = x_{n_j+1}$, or $w(j+1) = x_{n_j+1:m+1}$, where m is the smallest integer strictly larger than n_j such that $x_{n_j+1:m} \in \{w(1), \ldots, w(j)\}$ and $x_{n_j+1:m+1} \notin \{w(1), \ldots, w(j)\}$;
- v is empty or equal to one of the $w(j)$, $j \leqslant c(x_{1:n})$.

For instance, the decomposition of the following word containing 17 letters taken in the alphabet $\{0, 1, 2\}$

$$x_{1:17} = 00101021020210200$$

has 9 sequences

$$0, \ 01, \ 010, \ 2, \ 1, \ 02, \ 021, \ 020, \ 0$$

and we have $c(x_{1:17}) = 8$ and $v = 0$.

We then encode $w(j)$ by the integer i such that $w(j) = w(i)a$, with $a \in \mathcal{X}$, and by a. One has to indicate, in the code, when one refers to i and when one refers to a. The code is as follows:

$$f_{LZ}(x_{1:n}) = b(1) \cdot b(2) \ldots b(c) \cdot b(c+1)$$

with $c = c(x_{1:n})$, and

- If $j \leqslant c$ and $w(j) = a \in \mathcal{X}$, $b(j)$ is 0 followed by the binary code $g(a)$ of a with length $\lceil \log_2 |\mathcal{X}| \rceil$. For $j = 1$, one only needs $g(a)$ since we know that $w(1) \in \mathcal{X}$.
- If $j \leqslant c$ and $i < j$ is the smallest integer such that $w(j) = w(i)a$ where $a \in \mathcal{X}$, $b(j)$ is 1 followed by the binary code $h_j(i)$ of i, with $\lceil \log_2 j \rceil$ bits, since we know that $i < j$, and followed by $g(a)$.
- If $j = c + 1$, if v is empty, then $b(c+1)$ is empty too; otherwise $b(c+1)$ is $1h_{c+1}(i)$, where i is the smallest integer such that $v = w(i)$.

Decoding is obvious in view of the coding method: we start by considering the first $\lceil \log_2 |\mathcal{X}| \rceil$ symbols, which we decode in binary to obtain $w(1)$. Then we set

$j = 2$ and if the next symbol is 0, we decode $\lceil \log_2 |\mathscr{X}| \rceil$ symbols in binary to obtain $w(j)$, whereas if the next symbol is 1, we start by decoding the next $\lceil \log_2 j \rceil$ symbols in binary, which gives an integer $i < j$, and we decode the next $\lceil \log_2 |\mathscr{X}| \rceil$ symbols in binary to obtain a, and we set $w(j) = w(i)a$. We increase j by 1 and repeat. The entire decoded word is obtained by concatenation of the $w(j)$.

Clearly, both coding and decoding are made *online*.

Going back to our example, we have $\lceil \log_2 |\mathscr{X}| \rceil = 2$ and

$$b(1) = 00, \ b(2) = 1101, \ b(3) = 11000, \ b(4) = 010, \ b(5) = 001,$$

$$b(6) = 100110, \ b(7) = 111001, \ b(8) = 111000, \ b(9) = 10001.$$

Thus

$$f_{\mathrm{LZ}}(00101021020210200) = 001101110000100011001101110011110001001.$$

One may notice that in this example the length of the codeword is larger than the length of the encoded word. In fact, better compression is obtained when the encoded word is longer and when reference to already observed sequences becomes efficient.

The code constructed in this way is lossless, since it is decodable, and satisfies

$$\ell[f_{\mathrm{LZ}}(x_{1:n})]$$

$$\leqslant \lceil \log_2 |\mathscr{X}| \rceil + 1 + \sum_{j=1}^{c(x_{1:n})} \left(\lceil \log_2 j \rceil + 1 + \lceil \log_2 |\mathscr{X}| \rceil \right) + \lceil \log_2(c(x_{1:n}) + 1) \rceil + 1.$$

We thus have

$$\ell[f_{\mathrm{LZ}}(x_{1:n})] \leqslant (c(x_{1:n}) + 1)\left(\log_2 c(x_{1:n}) + \log_2 |\mathscr{X}| + 2 \right) + 1. \qquad (2.1)$$

The following theorem states that this code is weakly universal over the class of stationary ergodic processes with values in a finite alphabet.

Theorem 2.1 *If $\mathscr{X}$ is finite and if $(X_n)_{n \in \mathbb{N}}$ is a stationary ergodic process with law $\mathbb{P}$, then $\mathbb{P}$-a.s.*

$$\limsup_{n \to +\infty} \frac{\ell[f_{\mathrm{LZ}}(X_{1:n})]}{n} \leqslant H_*(\mathbb{P}).$$

The proof of this theorem only relies on the fact that the code is based on the decomposition of the word $x_{1:n}$ into $c(x_{1:n})$ pairwise distinct sequences and that the length of the encoded word is upper-bounded by a quantity asymptotic to $c(x_{1:n}) \log_2 c(x_{1:n})$. Hence, all code based on a decomposition of the word into pairwise distinct sequences and such that the length of the encoded word is upper-bounded by a quantity asymptotic to $c(x_{1:n}) \log_2 c(x_{1:n})$ is weakly universal over the class of stationary ergodic processes with values in a finite alphabet.

For all integers k, let $\mathbb{P}^k$ be the k-Markovian approximation of $\mathbb{P}$, as defined in (1.1) in Chap. 1.

Let $c = c(x_{1:n})$. For all $s \in \mathscr{X}^k$ and all integers m, denote by $c_{m,s}$ the number of $w(i)$ with length m and preceded by s in the word $x_{1:n}$ (if $w(i) = x_{n_i:n_{i+1}-1}$, we let $s = x_{n_i-k:n_i-1}$). Then

$$\sum_{\substack{s \in \mathscr{X}^k \\ m \in \mathbb{N}}} c_{m,s} = c, \qquad \sum_{\substack{s \in \mathscr{X}^k \\ m \in \mathbb{N}}} m c_{m,s} \leqslant n.$$

(We have $\sum_{s \in \mathscr{X}^k, m \in \mathbb{N}} m c_{m,s} = n$ if v is empty in the decomposition.) For all integers n and all words $x_{1:n}$ of length n, for all s in $\mathscr{X}^k$, we write $\mathbb{P}^k (x_{1:n} | s)$ instead of $\mathbb{P}^k (U_{k+1:k+n} = x_{1:n} | U_{1:k} = s)$, where $(U_n)_{n \geqslant 1}$ is a Markov chain of order k with law $\mathbb{P}^k$.

Lemma 2.2 (Ziv's Inequality) *For all integers k, for all $s_1 \in \mathscr{X}^k$,*

$$\log_2 \mathbb{P}^k (w(1) \ldots w(c) | s_1) \leqslant - \sum_{s,m} c_{m,s} \log_2 c_{m,s}.$$

Proof (Proof of Lemma 2.2). For $i = 1, \ldots, c$, denote by s_i the sequence of $\mathscr{X}^k$ which precedes $w(i)$ in the word $s_1 \cdot w(1) \ldots w(c)$. We then have

$$\log_2 \mathbb{P}^k (w(1) \ldots w(c) | s_1) = \sum_{i=1}^{c} \log_2 \mathbb{P}^k (w(i) | s_i)$$

$$= \sum_{m,s} c_{m,s} \left(\sum_{\substack{i:|w(i)|=m \\ s_i=s}} \frac{1}{c_{m,s}} \log_2 \mathbb{P}^k (w(i) | s_i) \right)$$

$$\leqslant \sum_{m,s} c_{m,s} \log_2 \left(\sum_{\substack{i:|w(i)|=m \\ s_i=s}} \frac{\mathbb{P}^k (w(i) | s_i)}{c_{m,s}} \right)$$

by Jensen's Inequality. But since the $w(i)$ are all distinct,

$$\sum_{\substack{i:|w(i)|=m \\ s_i=s}} \mathbb{P}^k (w(i) | s_i) \leqslant 1. \qquad \square$$

Theorem 2.3 *If $\mathscr{X}$ is finite, if $(X_n)_{n \in \mathbb{N}}$ is stationary ergodic with law $\mathbb{P}$ and finite entropy rate, if the decomposition procedure produces pairwise distinct sequences, then $\mathbb{P}$-a.s.*

$$\lim_{n \to +\infty} \frac{c(X_{1:n})}{n} = 0 \quad and \quad \limsup_{n \to +\infty} \frac{1}{n} c(X_{1:n}) \log_2 c(X_{1:n}) \leqslant H_*(\mathbb{P}).$$

Proof (Proof of Theorem 2.3). Let us fix n and $X_{1:n}$. Denote by (U, V) the random variable over $\mathbb{N} \times \mathscr{X}^k$ with law P such that $P\,(U = m, V = s) = c_{m,s}/c$, where we recall that $c = c(X_{1:n})$ and that $c_{m,s}$ is the number of $w(i)$ (in the decomposition of $X_{1:n}$) with length m and preceded by s in $X_{1:n}$. By Ziv's Inequality,

$$\log_2 \mathbb{P}^k\,(X_{1:n} \mid X_{-k+1:0}) \leqslant -c(X_{1:n}) \log_2 c(X_{1:n}) + c(X_{1:n}) H\,(U, V).$$

On the other hand,

$$E(U) = \frac{\sum_{s \in \mathscr{X}^k, m \in \mathbb{N}} m c_{m,s}}{c} \leqslant \frac{n}{c}.$$

Let us recall the following (proved at the end of the section), which allows us to upper-bound the entropy of a random variable, given an upper bound on its expectation.

Lemma 2.4 *If U is a random variable with values in $\mathbb{N}$ such that $E(U) \leqslant M$, then $H(U) \leqslant (M + 1) \log_2(M + 1) - M \log_2 M$.*

We have $H(V) \leqslant \log_2 |\mathscr{X}^k| = k \log_2 |\mathscr{X}|$. By Lemma 2.4, we obtain

$$H\,(U, V) \leqslant H\,(U) + H\,(V)$$
$$\leqslant k \log_2 |\mathscr{X}| + \left(\frac{n}{c} + 1\right) \log_2 \left(\frac{n}{c} + 1\right) - \frac{n}{c} \log_2 \frac{n}{c}.$$

It follows that

$$\frac{c}{n} \log_2 c \leqslant -\frac{1}{n} \log_2 \mathbb{P}^k\,(X_{1:n} \mid X_{-k+1:0})$$
$$+ \frac{c}{n} \left[k \log_2 |\mathscr{X}| + \left(\frac{n}{c} + 1\right) \log_2 \left(\frac{n}{c} + 1\right) - \frac{n}{c} \log_2 \frac{n}{c}\right]$$
$$\leqslant -\frac{1}{n} \log_2 \mathbb{P}^k\,(X_{1:n} \mid X_{-k+1:0})$$
$$+ \left(1 + \frac{c}{n}\right) \log_2 \left(1 + \frac{c}{n}\right) - \frac{c}{n} \log_2 \left(\frac{c}{n}\right) + \frac{c}{n} k \log_2 |\mathscr{X}|.$$

Let Ω be the event on which $\limsup_{n \to +\infty} \frac{c}{n} > 0$. On this event, $\lim_{n \to +\infty} c = +\infty$. Since $c/n \leqslant 1$, it follows that

$$0 < \limsup_{n \to +\infty} \frac{c}{n} \leqslant 1 \quad \text{and} \quad \limsup_{n \to +\infty} \frac{c}{n} \log_2 c = +\infty$$

by which we see that

$$\limsup_{n \to +\infty} -\frac{1}{n} \log_2 \mathbb{P}^k\,(X_{1:n} \mid X_{-k:-1}) = +\infty.$$

Now $\mathbb{P}$-a.s. $\lim_{n\to+\infty} -\frac{1}{n}\log_2 \mathbb{P}^k\left(X_{1:n}\mid X_{-k:-1}\right) = H^k$ is finite, thus $\mathbb{P}(\Omega) = 0$. Consequently, $\mathbb{P}$-a.s.

$$\lim_{n\to+\infty}\frac{c}{n} = 0,$$

establishing the first part of Theorem 2.3. For all integers k, $\mathbb{P}$-a.s.

$$\limsup_{n\to+\infty}\frac{1}{n}c(X_{1:n})\log_2 c(X_{1:n}) \leqslant H^k,$$

and the proof is concluded by taking the limit in k.

Proof (Proof of Theorem 2.1). Combine inequality (2.1) and Theorem 2.3.

Proof (Proof of Lemma 2.4). If U is a random variable with values in $\mathbb{N}$, let $\mu = E(U)$. Let W be a geometric random variable with expectation μ, that is

$$P(W = k) = \frac{1}{(\mu+1)}\left(\frac{\mu}{\mu+1}\right)^k.$$

Since the Kullback information between the law of U and the law of W is non-negative,

$$\sum_{k\geqslant 0}P(U = k)\log_2\frac{P(U = k)}{P(W = k)} \geqslant 0,$$

and thus

$$H(U) \leqslant -\sum_{k\geqslant 0}P(U = k)\log_2 P(W = k)$$

$$= \log_2(\mu+1) - \sum_{k\geqslant 0}kP(U = k)\log_2\frac{\mu}{(\mu+1)}$$

$$= (\mu+1)\log_2(\mu+1) - \mu\log_2\mu$$

because $\mu = E(U)$. But $(\mu+1)\log_2(\mu+1) - \mu\log_2\mu$ is increasing in $\mu > 0$, and $\mu \leqslant M$, which concludes the proof.

2.2   Strongly Universal Coding: Regrets and Redundancies

2.2.1   *Criteria*

Universal coding of stationary ergodic sources asymptotically leads to the minimal compression rate (the entropy rate) for any stationary ergodic source. However, this convergence might be slow: this is the price to pay for universality. One may nonetheless try to improve the performance on some classes of sources.

When the law of the source is known, one can code with a quasi-optimal compression rate by arithmetic coding. Is it possible to code as well as when the source is known, for a given set of sources called memoryless sources, which are sequences of i.i.d. random variables, or for the set of Markovian sources? One may think of estimating the law of the source *online*. To do so, one has to choose a model (a description of the chosen class of sources). One may use two classical statistical methods: maximum likelihood estimation, which leads to NML coding, and Bayesian estimation, relying on a *prior*, which leads to mixture coding.

Let $(X_n)_{n \in \mathbb{N}}$ be a source with law $\mathbb{P}$. Given an integer n, let Q_n be a probability over $\mathscr{X}^n$. The difference of code lengths between the code according to probability Q_n and the optimal ideal code for the sequence $x_{1:n}$ is

$$- \log_2 Q_n (x_{1:n}) + \log_2 \mathbb{P}_n (x_{1:n})$$

up to an additive term of at most 1 (since a code length has integer values).

Definition 2.2 The *redundancy* of Q_n is the expected difference of code lengths between the code according to probability Q_n and the optimal ideal code:

$$\bar{R}_n (Q_n, \mathbb{P}) = E_{\mathbb{P}} \left[- \log_2 Q_n (X_{1:n}) + \log_2 \mathbb{P} (X_{1:n}) \right] = D (\mathbb{P}_n \mid Q_n) ,$$

where $\mathbb{P}_n$ is the marginal law of $X_{1:n}$.

We know that if $\mathscr{X}$ is finite, it is possible to encode any stationary ergodic source in such a way that $\bar{R}_n (Q_n, \mathbb{P}) = o(n)$. Even more: almost surely, and not only in expectation, thanks to Theorem 2.1. We are now looking for optimality results for the speed.

Definition 2.3 Let $\mathscr{C}$ be a class of laws of processes over $\mathscr{X}^{\mathbb{N}}$. One says that $\rho(n)$ is a **weak speed** for $\mathscr{C}$ if $\rho(n) = o(n)$ and if there exists a sequence of codes over $\mathscr{X}^n$, thus a sequence of probabilities $(Q_n)_{n \in \mathbb{N}}$ over $(\mathscr{X}^n)_{n \in \mathbb{N}}$, such that for all $\mathbb{P}$ in $\mathscr{C}$, there exists a constant $K(\mathbb{P})$ such that

$$\bar{R}_n (Q_n, \mathbb{P}) \leqslant K(\mathbb{P})\rho(n).$$

Shields [2] showed that the class of stationary ergodic processes is too large.

Theorem 2.5 *There is no weak speed for the class of stationary ergodic processes.*

As in statistics, one may adopt either the minimax point of view or the Bayesian point of view to evaluate optimality over a class.

Definition 2.4 The *minimax redundancy* over class $\mathscr{C}$ is

$$\bar{R}_n (\mathscr{C}) = \inf_{Q_n} \sup_{\mathbb{P} \in \mathscr{C}} \bar{R}_n (Q_n, \mathbb{P}) ,$$

where the infimum is taken over probabilities over $\mathscr{X}^n$.

When computing $\overline{R}_n\,(Q_n, \mathbb{P})$, only the marginal law $\mathbb{P}_n$ of $\mathbb{P}$ over $\mathscr{X}^n$ intervenes. The class $\mathscr{C}_n$ of marginals over $\mathscr{X}^n$ of laws of $\mathscr{C}$ may be endowed with a sigma-field and a probability μ_n (this will be discussed in Sect. 2.2.3).

Definition 2.5 The *Bayesian redundancy* over class $\mathscr{C}$ is:

$$\underline{R}_n\,(\mathscr{C}) = \sup_{\mu_n}\inf_{Q_n}\int_{\mathscr{C}} \overline{R}_n\,(Q_n, \mathbb{P})\,\mathrm{d}\mu_n(\mathbb{P})$$

where the infimum is taken over probabilities over $\mathscr{X}^n$ and the supremum over probabilities over $\mathscr{C}_n$.

A more restrictive criterion, which adopts the trajectory (or *individual sequence*) point of view, is the following:

Definition 2.6 The *regret* over class $\mathscr{C}$ is:

$$R_n^*\,(\mathscr{C}) = \inf_{Q_n}\sup_{P\in\mathscr{C}}\sup_{x_{1:n}}\left\{-\log_2 Q_n\,(x_{1:n}) + \log_2 \mathbb{P}\,(x_{1:n})\right\}.$$

Clearly,

$$\underline{R}_n\,(\mathscr{C}) \leqslant \overline{R}_n\,(\mathscr{C}) \leqslant R_n^*\,(\mathscr{C})\ .$$

Note that all these criteria only depend on the class of marginals over $\mathscr{C}_n$.

2.2.2 NML, Regret and Redundancy

Let $\mathscr{C}_n$ be a class of laws over $\mathscr{X}^n$. When possible, we define the *normalized maximum likelihood probability*, denoted NML.

Definition 2.7 If $\sum_{y_{1:n}} \sup_{P\in\mathscr{C}_n} P(y_{1:n}) < +\infty$, the NML probability over $\mathscr{X}^n$ is defined as

$$\mathrm{NML}_n\,(x_{1:n}) = \frac{\sup_{P\in\mathscr{C}_n} P(x_{1:n})}{\sum_{y_{1:n}} \sup_{P\in\mathscr{C}_n} P(y_{1:n})}.$$

Note that in general, the quantity $\sup_{P\in\mathscr{C}_n} P(x_{1:n})$ depends on $x_{1:n}$.

Also note that $(\mathrm{NML}_n)_{n\geqslant 1}$ is not a consistent sequence of probability laws (NML_{n-1} is not the marginal, on the first $n-1$ coordinates, of NML_n).

Theorem 2.6 (cf. [3]) *The regret over class $\mathscr{C}_n$, whether it is finite or infinite, is always equal to*

$$R_n^*\,(\mathscr{C}_n) = \log_2\left(\sum_{y_{1:n}}\sup_{P\in\mathscr{C}_n} P(y_{1:n})\right).$$

When it is finite, the infimum in its definition is achieved by NML_n.

Proof (Proof of Theorem 2.6). Let

$$R_n(Q_n, \mathscr{C}_n)(x_{1:n}) = -\log_2 Q_n(x_{1:n}) + \log_2 \sup_{P \in \mathscr{C}_n} P(x_{1:n}).$$

Then

$$R_n^*(\mathscr{C}_n) = \inf_{Q_n} \sup_{x_{1:n}} R_n(Q_n, \mathscr{C}_n)(x_{1:n}).$$

Assume that $\sum_{y_{1:n}} \sup_{P \in \mathscr{C}_n} P(y_{1:n}) < +\infty$. Then NML_n is defined and

$$R_n(\mathrm{NML}_n, \mathscr{C}_n)(x_{1:n}) = \log_2 \left(\sum_{y_{1:n}} \sup_{P \in \mathscr{C}_n} P(y_{1:n}) \right)$$

does not depend on $x_{1:n}$. If $Q_n \neq \mathrm{NML}_n$, there exists an $x_{1:n}$ such that $Q_n(x_{1:n}) < \mathrm{NML}_n(x_{1:n})$ and thus

$$\sup_{y_{1:n}} R_n(Q_n, \mathscr{C}_n)(y_{1:n}) \geqslant R_n(Q_n, \mathscr{C}_n)(x_{1:n})$$

$$> R_n(\mathrm{NML}_n, \mathscr{C}_n)(x_{1:n}) = \sup_{y_{1:n}} R_n(\mathrm{NML}_n, \mathscr{C}_n)(y_{1:n}),$$

so that

$$R_n^*(\mathscr{C}_n) = R_n(\mathrm{NML}_n, \mathscr{C}_n)(x_{1:n}).$$

Now, if $R_n^*(\mathscr{C}_n)$ is finite, then there exists a Q_n such that for all $x_{1:n}$,

$$\sup_{P \in \mathscr{C}_n} P(x_{1:n}) \leqslant \left(1 + \exp R_n^*(\mathscr{C}_n)\right) Q_n(x_{1:n}).$$

Summing over $x_{1:n}$, we obtain

$$\sum_{x_{1:n}} \sup_{P \in \mathscr{C}_n} P(x_{1:n}) \leqslant 1 + \exp R_n^*(\mathscr{C}_n) < +\infty. \qquad \square$$

The connection between regret and redundancy of NML_n is simple:

$$D\left(\mathbb{P}_n \,|\, \mathrm{NML}_n\right) = E_{\mathbb{P}}\left[\log_2 \frac{\mathbb{P}(X_{1:n})}{\mathrm{NML}_n(X_{1:n})} \right]$$

$$= \log_2 \left(\sum_{y_{1:n}} \sup_{P \in \mathscr{C}_n} P(y_{1:n}) \right) + E_{\mathbb{P}}\left[\log_2 \frac{\mathbb{P}(X_{1:n})}{\sup_{P \in \mathscr{C}_n} P(X_{1:n})} \right]$$

$$= R_n^*(\mathscr{C}_n) + E_{\mathbb{P}}\left[\log_2 \frac{\mathbb{P}(X_{1:n})}{\sup_{P \in \mathscr{C}_n} P(X_{1:n})} \right].$$

Note that the last term of the inequality is not a Kullback divergence. In particular, this term is non-positive since the maximum likelihood is larger than the likelihood. Under weak regularity assumptions, in a parametric model, $2[\log_2 \sup_{P \in \mathscr{C}_n} P(X_{1:n}) - \log_2 \mathbb{P}(X_{1:n})]$ converges in distribution under $\mathbb{P}$ to a chi-square distribution, whose number of degrees of freedom is the dimension of the parameter space, see for instance [4, ch. 16]. If in addition equi-integrabilty holds, then the expectation converges to the dimension of the parameter space. For instance, when $\mathscr{X}$ has finite size k and when $\mathscr{C}_n$ is the class of marginals over $\mathscr{X}^n$ of memoryless sources, i.e. sequences of i.i.d. random variables, then the difference between redundancy of NML and regret converges to $-\frac{1}{2}(k-1)$.

One may precisely evaluate the regret for the class $\mathscr{C}$ of memoryless sources when $\mathscr{X} = \{1, \ldots, k\}$. Omitting $x_{1:n}$ from the notation, for $j = 1, \ldots, k$, let:

$$n_j = \sum_{i=1}^{n} 1_{x_i = j}.$$

Then, with the convention $0^0 = 1$:

$$\sup_{P \in \mathscr{C}} P(x_{1:n}) = \prod_{j=1}^{k} \left(\frac{n_j}{n}\right)^{n_j}.$$

Denote by $\mathscr{U}_\ell^n$ the set of partitions of size ℓ of the integer n, i.e. the set of ℓ-tuples of positive integers $(n_1, \ldots, n_\ell)$ such that $n_1 + \cdots + n_\ell = n$. We have

$$R_n^*(\mathscr{C}) = \log_2 \left[\sum_{\ell=1}^{k} C_k^\ell \sum_{(n_1, \ldots, n_\ell) \in \mathscr{U}_\ell^n} \frac{n!}{n_1! \ldots n_\ell!} \prod_{j=1}^{k} \left(\frac{n_j}{n}\right)^{n_j} \right],$$

which yields:

$$R_n^*(\mathscr{C}) = \frac{k-1}{2} \log_2 \frac{n}{2\pi} + \log_2 \frac{\Gamma(\frac{1}{2})^k}{\Gamma(\frac{k}{2})} + o(1). \tag{2.2}$$

Proof (Proof of (2.2)). Let us recall (see for instance [5]) that the Gamma function is defined over $\mathbb{C}$ by

$$\Gamma(z) = \int_0^\infty x^{z-1} e^{-x} dx, \quad \Gamma\left(\frac{1}{2}\right) = \sqrt{\pi}, \quad \Gamma(z+1) = z\Gamma(z), \tag{2.3}$$

and that for all real numbers $z > 0$, there exists a $\beta \in [0, 1]$ such that

$$\Gamma(z) = z^{z-\frac{1}{2}} e^{-z} e^{\frac{\beta}{12z}} \sqrt{2\pi}. \tag{2.4}$$

For all integers n, since $\Gamma(n+1) = n!$,

$$n! = \sqrt{2\pi n}\left(\frac{n}{e}\right)^n \varepsilon(n), \quad \varepsilon(n) = \left(1 + \frac{1}{n}\right)^{n+\frac{1}{2}} e^{-1} e^{\frac{\beta}{12(n+1)}}.$$

Since

$$e^{(\frac{1}{n} - \frac{1}{2n^2})(n+\frac{1}{2})} \leqslant \left(1 + \frac{1}{n}\right)^{n+\frac{1}{2}} \leqslant e^{1+\frac{1}{2n}}$$

we get

$$e^{-\frac{1}{4n^2}} \leqslant \varepsilon(n) \leqslant e^{\frac{7}{12n}},$$

and in particular $\varepsilon(n)$ tends to 1 as n tends to infinity. This implies

$$\sum_{(n_1,\dots,n_\ell)\in \mathscr{U}_\ell^n} \frac{n!}{n_1!\dots n_\ell!} \prod_{j=1}^{k} \left(\frac{n_j}{n}\right)^{n_j} = \left(\frac{1}{2\pi}\right)^{\frac{1}{2}(\ell-1)} \sum_{(n_1,\dots,n_\ell)\in \mathscr{U}_\ell^n} \sqrt{\frac{n}{n_1\dots n_\ell}} \cdot \frac{\varepsilon(n)}{\varepsilon(n_1)\dots\varepsilon(n_\ell)}$$

and thus

$$\left(\frac{1}{2\pi}\right)^{\frac{1}{2}(\ell-1)} \sum_{(n_1,\dots,n_\ell)\in \mathscr{U}_\ell^n} \sqrt{\frac{n}{n_1\dots n_\ell}} e^{-\frac{1}{4n^2} - \frac{7}{12}\sum_{i=1}^{\ell}\frac{1}{n_i}}$$

$$\leqslant \sum_{(n_1,\dots,n_\ell)\in \mathscr{U}_\ell^n} \frac{n!}{n_1!\dots n_\ell!} \prod_{j=1}^{k} \left(\frac{n_j}{n}\right)^{n_j}$$

$$\leqslant \left(\frac{1}{2\pi}\right)^{\frac{1}{2}(\ell-1)} \sum_{(n_1,\dots,n_\ell)\in \mathscr{U}_\ell^n} \sqrt{\frac{n}{n_1\dots n_\ell}} e^{\frac{7}{12n} + \frac{1}{4}\sum_{i=1}^{\ell}\frac{1}{n_i^2}}.$$

Let $\mathscr{S}_\ell$ be the simplex of $\mathbb{R}^\ell$:

$$\mathscr{S}_\ell = \left\{(x_1,\dots,x_\ell) \ : \ x_1 \geqslant 0,\dots,x_\ell \geqslant 0, \ \sum_{i=1}^{\ell} x_i = 1\right\}$$

and let us introduce the set

$$\mathbb{S}_\ell = \left\{(x_1,\dots,x_{\ell-1}) \ : \ x_1 \geqslant 0,\dots,x_{\ell-1} \geqslant 0, \ \sum_{i=1}^{\ell-1} x_i \leqslant 1\right\}.$$

When $(x_1,\dots,x_{\ell-1}) \in \mathbb{S}_\ell$, we set $x_\ell = 1 - \sum_{i=1}^{\ell-1} x_i$ so that $(x_1,\dots,x_\ell) \in \mathscr{S}_\ell$. For all $\varepsilon > 0$ let

$$\mathbb{S}_\ell^\varepsilon = \left\{(x_1,\dots,x_{\ell-1}) \ : \ x_1 \geqslant \varepsilon,\dots,x_{\ell-1} \geqslant \varepsilon, \ \sum_{i=1}^{\ell-1} x_i \leqslant 1-\varepsilon\right\}.$$

For $(n_1,\dots,n_\ell) \in \mathscr{U}_\ell^n$, using the hyper-rectangles

$$\frac{n_1 - 1}{n} \leqslant x_1 \leqslant \frac{n_1}{n}, \ldots, \frac{n_{\ell-1} - 1}{n} \leqslant x_{\ell-1} \leqslant \frac{n_{\ell-1}}{n}$$

we obtain, for all $\varepsilon > 0$,

$$\left(\frac{1}{2\pi}\right)^{\frac{1}{2}(\ell-1)} \sum_{\substack{(n_1,\ldots,n_\ell)\in\mathscr{U}_\ell^n \\ n_1 \geqslant n\varepsilon,\ldots,n_\ell \geqslant n\varepsilon}} \sqrt{\frac{n}{n_1 \ldots n_\ell}} \, e^{\frac{7}{12n} + \frac{1}{4}\sum_{i=1}^{\ell} \frac{1}{n_i^2}}$$

$$\leqslant \sqrt{1 + \frac{\ell-1}{n\varepsilon}} \, e^{\frac{7}{12n} + \frac{\ell}{4\varepsilon^2 n^2}} \left(\frac{n}{2\pi}\right)^{\frac{1}{2}(\ell-1)} \int_{\mathbb{S}_\ell} \frac{dx_1 \ldots dx_{\ell-1}}{\sqrt{x_1 \ldots x_\ell}},$$

and

$$\left(\frac{1}{2\pi}\right)^{\frac{1}{2}(\ell-1)} \sum_{\substack{(n_1,\ldots,n_\ell)\in\mathscr{U}_\ell^n \\ \exists i:n_i \leqslant n\varepsilon}} \sqrt{\frac{n}{n_1 \ldots n_\ell}} \, e^{\frac{7}{12n} + \frac{1}{4}\sum_{i=1}^{\ell} \frac{1}{n_i^2}}$$

$$\leqslant \sqrt{\ell} \, e^{1 + \frac{\ell}{4}} \left(\frac{n}{2\pi}\right)^{\frac{1}{2}(\ell-1)} \int_{\mathbb{S}_\ell \cap \{\exists i:x_i \leqslant \varepsilon\}} \frac{dx_1 \ldots dx_{\ell-1}}{\sqrt{x_1 \ldots x_\ell}}$$

$$\leqslant \left(\frac{n}{2\pi}\right)^{\frac{1}{2}(\ell-1)} C\sqrt{\varepsilon}$$

for some constant $C > 0$. Hence

$$\left(\frac{1}{2\pi}\right)^{\frac{1}{2}(\ell-1)} \sum_{(n_1,\ldots,n_\ell)\in\mathscr{U}_\ell^n} \sqrt{\frac{n}{n_1 \ldots n_\ell}} \, e^{\frac{7}{12n} + \frac{1}{4}\sum_{i=1}^{\ell} \frac{1}{n_i^2}}$$

$$\leqslant \left(\frac{n}{2\pi}\right)^{\frac{1}{2}(\ell-1)} \left[\int_{\mathbb{S}_\ell} \frac{dx_1 \ldots dx_{\ell-1}}{\sqrt{x_1 \ldots x_\ell}} + O\left(\sqrt{\varepsilon}\right) + o\left(1\right)\right]$$

and

$$\left(\frac{1}{2\pi}\right)^{\frac{1}{2}(\ell-1)} \sum_{(n_1,\ldots,n_\ell)\in\mathscr{U}_\ell^n} \sqrt{\frac{n}{n_1 \ldots n_\ell}} \, e^{\frac{7}{12n} + \frac{1}{4}\sum_{i=1}^{\ell} \frac{1}{n_i^2}}$$

$$\leqslant \left(\frac{n}{2\pi}\right)^{\frac{1}{2}(\ell-1)} \left[\int_{\mathbb{S}_\ell} \frac{dx_1 \ldots dx_{\ell-1}}{\sqrt{x_1 \ldots x_\ell}} + o\left(1\right)\right].$$

Using now the hyper-rectangles

$$\frac{n_1}{n} \leqslant x_1 \leqslant \frac{n_1 + 1}{n}, \ldots, \frac{n_{\ell-1}}{n} \leqslant x_{\ell-1} \leqslant \frac{n_{\ell-1} + 1}{n}$$

we obtain

$$\left(\frac{1}{2\pi}\right)^{\frac{1}{2}(\ell-1)} \sum_{(n_1,\dots,n_\ell)\in\mathscr{U}_\ell^n} \sqrt{\frac{n}{n_1\dots n_\ell}}\, e^{-\frac{1}{4n^2}-\frac{7}{12}\sum_{i=1}^{\ell}\frac{1}{n_i}}$$

$$\geq \left(\frac{1}{2\pi}\right)^{\frac{1}{2}(\ell-1)} \sum_{\substack{(n_1,\dots,n_\ell)\in\mathscr{U}_\ell^n \\ n_1\geq n\varepsilon,\dots,n_\ell\geq n\varepsilon}} \sqrt{\frac{n}{n_1\dots n_\ell}}\, e^{-\frac{1}{4n^2}-\frac{7\ell}{12n\varepsilon}}$$

$$\geq e^{-\frac{1}{4n^2}-\frac{7\ell}{12n\varepsilon}} \left(\frac{n}{2\pi}\right)^{\frac{1}{2}(\ell-1)} \int_{\mathbb{S}_\ell^\varepsilon} \frac{dx_1\dots dx_{\ell-1}}{\sqrt{x_1\dots x_\ell}}.$$

But when ε tends to 0, $\int_{\mathbb{S}_\ell^\varepsilon}\frac{dx_1\dots dx_{\ell-1}}{\sqrt{x_1\dots x_\ell}}$ tends to $\int_{\mathbb{S}_\ell}\frac{dx_1\dots dx_{\ell-1}}{\sqrt{x_1\dots x_\ell}}$, and altogether

$$\sum_{(n_1,\dots,n_\ell)\in\mathscr{U}_\ell^n} \frac{n!}{n_1!\dots n_\ell!} \prod_{j=1}^{k}\left(\frac{n_j}{n}\right)^{n_j} = \left(\frac{n}{2\pi}\right)^{\frac{1}{2}(\ell-1)} \left[\int_{\mathbb{S}_\ell}\frac{dx_1\dots dx_{\ell-1}}{\sqrt{x_1\dots x_\ell}} + o(1)\right].$$

Now (we will come back to this in Sect. 2.4), a simple change of variables shows that

$$\Gamma\left(\tfrac{1}{2}\right)^\ell = \Gamma\left(\tfrac{1}{2}\ell\right) \int_{\mathbb{S}_\ell} \frac{dx_1\dots dx_{\ell-1}}{\sqrt{x_1\dots x_\ell}},$$

so that

$$R_n^*(\mathscr{C}) = \log_2\left[\sum_{\ell=1}^{k} C_k^\ell \left(\frac{n}{2\pi}\right)^{\frac{1}{2}(\ell-1)} \left\{\Gamma\left(\tfrac{1}{2}\right)^\ell / \Gamma\left(\tfrac{1}{2}\ell\right) + o(1)\right\}\right]$$

and we see that the first term ($\ell = k$) is dominant.

2.2.3 Minimax and Maximin

The framework here will be very general and we will show that Bayesian redundancy and minimax redundancy are equal. One may refer to Dudley's book [6] for the topological concepts.

We do not assume here that $\mathscr{X}$ is necessarily finite or countable, but that it is a complete separable metric space, endowed with the Borel sigma-field. Its metric is denoted by $d_{\mathscr{X}}$.

If f is a real-valued function over $\mathscr{X}$, let

$$\|f\|_{BL} = \sup_{x\in\mathscr{X}}|f(x)| + \sup_{x\neq y}\frac{|f(x)-f(y)|}{d_{\mathscr{X}}(x,y)}.$$

The set of probability measures over $\mathscr{X}$ is denoted by $\mathscr{P}(\mathscr{X})$ and is endowed with the distance

$$\rho(P, Q) = \sup \left\{ \left| \int f\,\mathrm{d}P - \int f\,\mathrm{d}Q \right| : \|f\|_{\mathrm{BL}} \leqslant 1 \right\}.$$

The set $\mathscr{P}(\mathscr{X})$, endowed with this distance, is a complete separable metric space. It is endowed with its Borel sigma-field. If (P_n) is a sequence of probability measures over $\mathscr{X}$ and P is a probability measure over $\mathscr{X}$, the probability measure P_n converges weakly to P if and only if $\rho(P_n, P) \to 0$.

Let $\mathscr{C}$ a Borel subset of $\mathscr{P}(\mathscr{X})$. One may also define the set $\mathscr{P}(\mathscr{C})$ of probability measures over $\mathscr{C}$, endowed with the weak convergence topology. For all $\mu \in \mathscr{P}(\mathscr{C})$, we denote by P_μ the probability measure over $\mathscr{X}$ such that for all Borel subsets A of $\mathscr{X}$,

$$P_\mu(A) = \int P(A)\,\mathrm{d}\mu(P).$$

We will repeatedly use the fact that all probabilities P over a complete separable metric space are tight, i.e. for all $\varepsilon > 0$, there exists a compact set K such that $P(K) \geqslant 1 - \varepsilon$.

Let us start with general considerations on Kullback information and then on Bayesian redundancy.

Definition 2.8 The *Kullback information* (or *relative entropy*) $D(P \mid Q)$ between two probabilities P and Q over $\mathscr{X}$ is defined by

$$D(P \mid Q) = \begin{cases} \displaystyle\int \left(\log_2 \frac{\mathrm{d}P}{\mathrm{d}Q} \right)\mathrm{d}P & \text{if } P \text{ is absolutely continuous with respect to } Q, \\ +\infty & \text{otherwise.} \end{cases}$$

This definition makes sense and generalizes Definition 1.4. Indeed, if it is well-defined, we have

$$\int \left(\log_2 \frac{\mathrm{d}P}{\mathrm{d}Q} \right)\mathrm{d}P = \int \left(\log_2 \frac{\mathrm{d}P}{\mathrm{d}Q} \right)_+ \mathrm{d}P - \int \left(\log_2 \frac{\mathrm{d}P}{\mathrm{d}Q} \right)_- \mathrm{d}P, \qquad (2.5)$$

where $\left(\log_2 \mathrm{d}P/\mathrm{d}Q \right)_+$ are $\left(\log_2 \mathrm{d}P/\mathrm{d}Q \right)_-$ are respectively the positive and negative parts of $\log_2 \mathrm{d}P/\mathrm{d}Q$. Since $u \mapsto u \log_2 u$ is lower-bounded by a constant $-C$, with $C > 0$, for $u \in\,]0, 1]$, $-\int \left(\log_2 \mathrm{d}P/\mathrm{d}Q \right)_- \mathrm{d}P$ is always finite and between $-C$ and 0, so that Eq. (2.5) defines an integral, possibly equal to $+\infty$ if $\int \left(\log_2 \mathrm{d}P/\mathrm{d}Q \right)_+ \mathrm{d}P = +\infty$.

Moreover, $u \mapsto -\log_2 u$ being a strictly convex function, Jensen's Inequality gives $D(P \mid Q) \geqslant 0$ and $D(P \mid Q) = 0$ if and only if $P = Q$.

We can also generalize Definition 1.5 of mutual information.

Definition 2.9 Let $\mathscr{X}_1$ and $\mathscr{X}_2$ two complete separable metric spaces and (W, X) a random variable with law P over $\mathscr{X}_1 \times \mathscr{X}_2$, W having marginal distribution P_W over $\mathscr{X}_1$ and X having marginal distribution P_X over $\mathscr{X}_2$. The **mutual information** of W and X is the Kullback information between the joint distribution and the product of marginal distributions:

$$I(W; X) = D(P \mid P_W \otimes P_X).$$

Let $\mathscr{F}_b$ be the set of bounded measurable functions over $\mathscr{X}$ and $\mathscr{F}_c$ be the set of bounded continuous functions over $\mathscr{X}$. Let also

$$\underline{R}(\mathscr{C}) = \sup_{\mu \in \mathscr{P}(\mathscr{C})} \inf_{Q \in \mathscr{P}(\mathscr{X})} \int D(P \mid Q) \, d\mu(P),$$

$$\overline{R}(\mathscr{C}) = \inf_{Q \in \mathscr{P}(\mathscr{X})} \sup_{P \in \mathscr{C}} D(P \mid Q).$$

Proposition 2.7 *For all $f \in \mathscr{F}_b$ and for all $Q \in \mathscr{P}(\mathscr{X})$,*

$$-\log_2 \int 2^f dQ = \min_{P \in \mathscr{P}(\mathscr{X})} \left\{ D(P \mid Q) - \int f dP \right\}.$$

Proof (Proof of Proposition 2.7). Let f be an element of $\mathscr{F}_b$ and P_f the probability $\frac{2^f}{\int 2^f dQ} Q$, which is absolutely continuous with respect to Q. The infimum in Proposition 2.7 is finite since $D(P_f \mid Q) = \int f dP_f - \log_2 \int 2^f dQ$. Let now P be such that $D(P \mid Q) < +\infty$. P is absolutely continuous with respect to Q, and $dP/dQ = \frac{2^f}{\int 2^f dQ} dP/dP_f$ so that $D(P \mid Q) = D(P \mid P_f) + \int f dP - \log_2 \int 2^f dQ$. Hence

$$-\log_2 \int 2^f dQ \leqslant D(P \mid Q) - \int f dP,$$

and there is equality for $P = P_f$.

We may then characterize the relative entropy thanks to the following theorem.

Theorem 2.8 *For all P and Q in $\mathscr{P}(\mathscr{X})$:*

$$D(P \mid Q) = \sup_{f \in \mathscr{F}_b} \left\{ \int f dP - \log_2 \int 2^f dQ \right\}$$

$$= \sup_{f \in \mathscr{F}_c} \left\{ \int f dP - \log_2 \int 2^f dQ \right\}.$$

Proof (Proof of Theorem 2.8). By Proposition 2.7,

$$D(P \mid Q) \geqslant \sup_{f \in \mathscr{F}_b} \left\{ \int f dP - \log_2 \int 2^f dQ \right\}.$$

Assume that this supremum is finite (otherwise there is nothing to prove) and denote it by S. Let us first show that P is absolutely continuous with respect to Q. Let A be such that $Q(A) = 0$. Then for all $r > 0$, taking $f(x) = r1_A(x)$, it follows that $rP(A) \leqslant S$, and thus $P(A) = 0$ by letting $r \to +\infty$.

Now, for all $t \in [0, 1]$ and all integers n, let $f_{t,n} = (t + (1 - t)\mathrm{d}P/\mathrm{d}Q) \wedge n$. Then $f_{t,n}$ belongs to $\mathscr{F}_b$ and when n tends to infinity, $f_{t,n}$ increases to $f_t = t + (1 - t)\mathrm{d}P/\mathrm{d}Q$. By monotone convergence:

$$S \geqslant \int \log_2 f_t \mathrm{d}P - \log_2 \int f_t \mathrm{d}Q = \int \log_2 f_t \mathrm{d}P.$$

By concavity of the function $u \mapsto \log_2 u$, we have $\log_2 f_t \geqslant (1 - t)\log_2 \frac{\mathrm{d}P}{\mathrm{d}Q}$, thus $S \geqslant (1 - t)D(P\,|\,Q)$, and we obtain $S \geqslant D(P\,|\,Q)$ by letting t tend to 0.

The first equality of the theorem follows.

Since $\mathscr{F}_c \subset \mathscr{F}_b$,

$$\sup_{f \in \mathscr{F}_c} \left\{ \int f \mathrm{d}P - \log_2 \int 2^f \mathrm{d}Q \right\} \leqslant \sup_{f \in \mathscr{F}_b} \left\{ \int f \mathrm{d}P - \log_2 \int 2^f \mathrm{d}Q \right\}.$$

Let us now establish the reverse inequality. Let $\varepsilon > 0$. The probability measures P and Q are tight, so there exists a compact subset K of $\mathscr{X}$ such that $P(K^c) \leqslant \varepsilon$ and $Q(K^c) \leqslant \varepsilon$. By Lusin's Theorem and the Tietze-Urysohn Extension Theorem (see Dudley [6]), there exists a closed set $F \subset K$ and a function $g \in \mathscr{F}_c$ such that $(P + Q)(F^c) \leqslant \varepsilon$, f and g coincide on F and $\|g\|_\infty \leqslant \|f\|_\infty$. We thus have

$$\int f \mathrm{d}P - \log_2 \int 2^f \mathrm{d}Q = \int g \mathrm{d}P - \log_2 \int 2^g \mathrm{d}Q + \int_{F^c} f \mathrm{d}P$$

$$- \log_2 \int_{F^c} 2^f \mathrm{d}Q - \int_{F^c} g \mathrm{d}P + \log_2 \int_{F^c} 2^g \mathrm{d}Q$$

$$\leqslant \int g \mathrm{d}P - \log_2 \int 2^g \mathrm{d}Q + 4\varepsilon \|f\|_\infty$$

$$\leqslant \sup_{h \in \mathscr{F}_c} \left\{ \int h \mathrm{d}P - \log_2 \int 2^h \mathrm{d}Q \right\} + 4\varepsilon \|f\|_\infty.$$

Letting ε tend to 0 and taking the supremum in $f \in \mathscr{F}_b$, we obtain the desired inequality, which concludes the proof.

Corollary 2.9 *For all partitions $B_1, \ldots, B_n$ of $\mathscr{X}$,*

$$D(P\,|\,Q) \geqslant \sum_{i=1}^n P(B_i) \log_2 \frac{P(B_i)}{Q(B_i)}.$$

Proof (Proof of Corollary 2.9). Apply Theorem 2.8 to the function

$$f(x) = \sum_{i=1}^{n} 1_{B_i}(x) \log_2 \frac{P(B_i)}{Q(B_i)}. \qquad \square$$

Corollary 2.10 *For all* $P, Q \in \mathscr{P}(\mathscr{X})$,

- $D(P \,|\, \cdot)$ *is convex lower-semicontinuous;*
- $D(\cdot \,|\, Q)$ *is convex lower-semicontinuous;*
- $D(\cdot \,|\, \cdot)$ *is convex lower-semicontinuous;*
- $\mu \mapsto \inf_Q \int D(P \,|\, Q) \mathrm{d}\mu(P)$ *is concave.*

Proof By Theorem 2.8, $D(P \,|\, \cdot)$, $D(\cdot \,|\, Q)$ and $D(\cdot \,|\, \cdot)$ are suprema of continuous convex functions. For all $Q \in \mathscr{P}(\mathscr{X})$, the function $\mu \mapsto \int D(P \,|\, Q) \mathrm{d}\mu(P)$ is linear, so the infimum of those functions is concave.

Note that by the Portmanteau Theorem, for all $Q \in \mathscr{P}(\mathscr{X})$, the function $\mu \mapsto \int D(P \,|\, Q) \mathrm{d}\mu(P)$ is lower-semicontinuous, but the infimum of lower-semicontinuous functions does not have any particular regularity, so $\mu \mapsto \inf_Q \int D(P \,|\, Q) \mathrm{d}\mu(P)$ is concave without any particular regularity.

Proposition 2.11 *For all* $\mu \in \mathscr{P}(\mathscr{C})$,

$$\inf_{Q \in \mathscr{P}(\mathscr{X})} \int D(P \,|\, Q) \, \mathrm{d}\mu(P) = \int D(P \,|\, P_\mu) \, \mathrm{d}\mu(P)$$

and P_μ *is the unique distribution achieving the infimum. This infimum is equal to* $I(W; X)$, *where* W *has distribution* μ *over* $\mathscr{C}$ *and the conditional distribution of* X *given* W *is* W.

Proof (Proof of Proposition 2.11). If for all $Q \in \mathscr{P}(\mathscr{X})$, $\int D(P \,|\, Q) \, \mathrm{d}\mu(P)$ is infinite, then the desired equality holds. Otherwise, for all $Q \in \mathscr{P}(\mathscr{X})$ such that $\int D(P \,|\, Q) \, \mathrm{d}\mu(P)$ is finite, by convexity of $D(\cdot \,|\, Q)$, we have $\int D(P \,|\, Q) \, \mathrm{d}\mu(P) \geqslant D(P_\mu \,|\, Q)$. Hence $D(P_\mu \,|\, Q)$ is finite and we have

$$\int D(P \,|\, Q) \, \mathrm{d}\mu(P) - D(P_\mu \,|\, Q) = E_\mu E_P \log_2 \frac{\mathrm{dP}}{\mathrm{dQ}}(X) - E_\mu E_P \log_2 \frac{\mathrm{dP}_\mu}{\mathrm{dQ}}(X)$$

$$= E_\mu E_P \log_2 \frac{\mathrm{dP}}{\mathrm{dP}_\mu}(X) = \int D(P \,|\, P_\mu) \, \mathrm{d}\mu(P).$$

The proposition follows, and we have

$$I(W; X) = D(\mu P \,|\, \mu \otimes P_\mu) = \int D(P \,|\, P_\mu) \, \mathrm{d}\mu(P). \qquad \square$$

Let $\overline{\mathscr{C}}$ be the topological closure of the set $\mathscr{C}$. Let us recall that a set is precompact if for all $\varepsilon > 0$, one can cover it by a finite union of balls of radius ε.

Theorem 2.12 *We always have*

$$\underline{R}(\mathscr{C}) = \overline{R}(\mathscr{C}).$$

- *If $\mathscr{C}$ is not precompact, this quantity is infinite.*
- *If $\mathscr{C}$ is precompact and if $\underline{R}(\mathscr{C})$ is finite, there exists a distribution $\hat{\mu}$ over $\overline{\mathscr{C}}$ such that $P_{\hat{\mu}}$ is a minimax coding distribution, i.e. it achieves $\overline{R}(\mathscr{C})$. If furthermore there exists a distribution $\hat{\mu}$ over $\overline{\mathscr{C}}$ achieving the maximin $\underline{R}(\mathscr{C})$, then $P_{\hat{\mu}}$ is also a minimax coding distribution.*

Remark 2.10 The equality between minimax redundancy and Bayesian redundancy, when $\mathscr{C}$ is compact, follows from Sion's Theorem [7]. The first two statements after the equality in the theorem can be found in Haussler [8], as well as the proof of the first of these statements (namely, that this quantity is infinite whenever $\mathscr{C}$ is not precompact).

Csiszár [9] considers the case where $\mathscr{X}$ is finite and shows that there exists a unique distribution μ over $\mathscr{C}$ such that μ is maximin.

It may happen that $\underline{R}(\mathscr{C})$ is infinite whereas $\mathscr{C}$ is precompact, see Proposition 3.3.

Proof (Proof of Theorem 2.12). Let us first consider the case where $\mathscr{C}$ is not precompact. We will show that $\underline{R}(\mathscr{C})$ is infinite. Since it always holds that $\overline{R}(\mathscr{C}) \geqslant \underline{R}(\mathscr{C})$, this will establish equality in that case.

The set $\mathscr{C}$ is not tight, meaning that there exists an $\varepsilon > 0$ such that for all compact subsets K of $\mathscr{X}$, there exists a P in $\mathscr{C}$ such that $P(K) \leqslant 1 - \varepsilon$ (see Dudley [6, Theorem 11.5.4]). Let $\delta = \frac{1}{2}\varepsilon$. Let n be some integer. One can find $P_1, \ldots, P_n$ in $\mathscr{C}$ and disjoint Borel sets $B_1, \ldots, B_n$ of $\mathscr{X}$ such that $A_n = \bigcup_{i=1}^n B_i$ is compact and $P_i(B_i) \geqslant \delta$ for all i.

The construction is done by induction. For $k = 1$, we take $P_1 \in \mathscr{C}$ and since P_1 is tight, there exists a compact set B_1 of $\mathscr{X}$ and P_1 such that $P_1(B_1) \geqslant \delta$. For $k \geqslant 1$, assume that we found $P_1, \ldots, P_k$ in $\mathscr{C}$ and disjoint Borel sets $B_1, \ldots, B_k$ of $\mathscr{X}$ such that $A_k = \bigcup_{i=1}^k B_i$ is compact and $P_i(B_i) \geqslant \delta$ for $i = 1, \ldots, k$. Since $\mathscr{C}$ is not tight, there exists a P_{k+1} in $\mathscr{C}$ such that $P_{k+1}(A_k) \leqslant 1 - \varepsilon$. Let now K be a compact set such that $P_{k+1}(K) \geqslant 1 - \delta$. Let $A_{k+1} = K \cup A_k$, which is compact, and B_{k+1} the complement of A_k in A_{k+1}: $B_{k+1} = A_{k+1} \setminus A_k$. Then $P_{k+1}(B_{k+1}) \geqslant \varepsilon - \delta = \delta$.

We now set $Q = \frac{1}{n} \sum_{i=1}^n P_i$. We have

$$\underline{R}(\mathscr{C}) \geqslant \frac{1}{n} \sum_{i=1}^n D(P_i \,|\, Q).$$

Denote by B_{n+1} the Borel set such that $(B_i)_{i=1,\ldots,n+1}$ form a partition of $\mathscr{X}$. Then

$$\frac{1}{n}\sum_{i=1}^{n} D\left(P_i \mid Q\right) \geqslant \frac{1}{n}\sum_{i=1}^{n}\sum_{j=1}^{n+1} P_i(B_j)\log_2 \frac{P_i(B_j)}{Q(B_j)}$$

$$\geqslant -\frac{\log_2 e}{e} + \frac{1}{n}\sum_{i=1}^{n} P_i(B_i)\log_2 \frac{P_i(B_i)}{Q(B_i)}$$

$$\geqslant -\frac{\log_2 e}{e} + \frac{1}{n}\left(\frac{\sum_{i=1}^{n} P_i(B_i)}{\sum_{i=1}^{n} Q(B_i)}\right)\log_2\left(\frac{\sum_{i=1}^{n} P_i(B_i)}{\sum_{i=1}^{n} Q(B_i)}\right)$$

$$\geqslant -\frac{\log_2 e}{e} + \delta \log_2 n\delta$$

for n large enough. The first inequality comes from Corollary 2.9. The second inequality comes from $x \log_2 x \geqslant -\log_2 e/e$ applied to $x = P_i(B_j)/Q(B_j)$, yielding

$$\frac{1}{n}\sum_{i=1}^{n}\sum_{j\neq i} P_i(B_j)\log_2 \frac{P_i(B_j)}{Q(B_j)} \geqslant \left(-\frac{\log_2 e}{e}\right)\frac{1}{n}\sum_{i=1}^{n}\sum_{j\neq i} Q(B_j) \geqslant -\frac{\log_2 e}{e}.$$

The third inequality follows from Jensen's Inequality, and the last uses the fact that the function $x \log_2 x$ is increasing on $[1, +\infty[$ and that $\left(\sum_{i=1}^{n} P_i(B_i)\right) / \left(\sum_{i=1}^{n} Q(B_i)\right) \geqslant n\delta \geqslant 1$ for n large enough. We obtain, for n large enough,

$$\underline{R}\left(\mathscr{C}\right) \geqslant -\frac{\log_2 e}{e} + \delta \log_2 n\delta,$$

and thus $\underline{R}\left(\mathscr{C}\right) = +\infty$.

Let us now consider the case where $\mathscr{C}$ is precompact. If $\underline{R}\left(\mathscr{C}\right) = +\infty$, then $\overline{R}\left(\mathscr{C}\right) = +\infty$. We may thus assume that $\underline{R}\left(\mathscr{C}\right) = M$ is finite. Let μ_n in $\mathscr{P}(\mathscr{C})$ such that

$$\int D\left(P \mid P_{\mu_n}\right) d\mu_n\left(P\right) \geqslant M - \frac{1}{n}.$$

Let P_0 in $\mathscr{C}$. For all $t \in]0, 1[$, let $\mu_{t,n} = (1 - t)\mu_n + t\delta_{P_0}$. Then, by Proposition 2.11,

$$\int D\left(P \mid P_{\mu_{t,n}}\right) d\mu_{t,n}\left(P\right) \leqslant M \leqslant \int D\left(P \mid P_{\mu_n}\right) d\mu_n\left(P\right) + \frac{1}{n}.$$

But

$$\int D\left(P \mid P_{\mu_{t,n}}\right) d\mu_{t,n}\left(P\right) = t D\left(P_0 \mid P_{\mu_{t,n}}\right) + (1 - t)\int D\left(P \mid P_{\mu_{t,n}}\right) d\mu_n$$

and $\int D\left(P \mid P_{\mu_{t,n}}\right) d\mu_n \geqslant \int D\left(P \mid P_{\mu_n}\right) d\mu_n$ by Proposition 2.11. Hence

$$t D\left(P_0 \mid P_{\mu_{t,n}}\right) \leqslant t\int D\left(P \mid P_{\mu_n}\right) d\mu_n + \frac{1}{n},$$

and

$$D\left(P_0 \mid P_{\mu_{t,n}}\right) \leqslant M + \frac{1}{tn}. \tag{2.6}$$

As $\mathscr{C}$ is precompact, $\overline{\mathscr{P}(\mathscr{C})} = \mathscr{P}(\overline{\mathscr{C}})$ is compact, and the sequence $(\mu_n)_n$ has an accumulation point $\overline{\mu}$. Thus, the sequence $(\mu_{t,n})_n$ has an accumulation point $\overline{\mu}_t = t\delta_{P_0} + (1-t)\overline{\mu}$. Since $\mu \mapsto P_\mu$ is a continuous map from $\mathscr{P}(\mathscr{C})$ to $\mathscr{P}(\mathscr{X})$ and since $D(P_0 \mid .)$ is lower-semicontinuous, (2.6) entails:

$$\forall P_0 \in \mathscr{C}, \quad D\left(P_0 \mid P_{\overline{\mu}_t}\right) \leqslant \underline{R}(\mathscr{C}).$$

Taking the limit as $t \to 0$ and by lower-semicontinuity, we get

$$\forall P_0 \in \mathscr{C}, \quad D\left(P_0 \mid P_{\overline{\mu}}\right) \leqslant \underline{R}(\mathscr{C}),$$

thus $\overline{R}(\mathscr{C}) = \underline{R}(\mathscr{C})$ and $P_{\overline{\mu}}$ is minimax.

Finally, if $\hat{\mu}$ is a distribution over $\overline{\mathscr{C}}$ achieving the maximin, i.e. such that

$$\underline{R}(\mathscr{C}) = \int D\left(P \mid P_{\hat{\mu}}\right) \mathrm{d}\hat{\mu}(P),$$

we may take $\mu_n = \hat{\mu}$ for all n in the previous proof, thus $\overline{\mu} = \hat{\mu}$ and $P_{\hat{\mu}}$ achieves the minimax.

Remark 2.11 This theorem gives a method to lower bound the minimax redundancy. Indeed, since

$$\overline{R}(\mathscr{C}) = \sup_W I(W; X),$$

where the supremum is taken over random variables W over $\mathscr{C}$, and where the conditional distribution of X given W is W, we get a lower bound by a judicious choice of W. This method will be used for renewal sources, see Sect. 2.5; for memoryless sources with values in an infinite alphabet, see Chap. 3.

2.3 Bayesian Redundancy

2.3.1 Rissanen's Theorem

We keep our general framework where $\mathscr{X}$ is a complete separable metric space, not necessarily finite or countable. We state a theorem giving a lower bound on the redundancy. The first theorem of this kind was proved by Rissanen [10] and was then improved and extended. The version presented here is due to Barron and Hengartner [11], and holds in a non-parametric setting (with Θ in a metric space of possibly infinite dimension). As a consequence of this lower bound on the redundancy, one obtains an asymptotic lower bound on the Bayesian redundancy, as well as an asymptotic lower bound on the statistical minimax risk when the loss function is the Kullback divergence.

Let $(X_n)_{n \in \mathbb{N}}$ be a sequence of random variables over $\mathscr{X}$, P_θ^n be the distribution of $X_{1:n}$, $\theta \in \Theta$, with $\Theta \subset \mathbb{R}^k$.

Theorem 2.13 *Assume that for all bounded subsets K of $\mathbb{R}^k$, there exists a sequence of estimators $\hat{\theta}_n$ such that Lebesgue-almost everywhere in θ, $\sqrt{n}(\hat{\theta}_n - \theta)$ is tight under P_θ^n. Then, for all sequence of distributions Q_n over $\mathscr{X}^n$, Lebesgue-almost everywhere in θ,*

$$\limsup_{n \to +\infty} \frac{D\left(P_\theta^n \mid Q_n\right)}{\log_2 n} \geqslant \frac{k}{2}.$$

Moreover, if K is a subset of $\mathbb{R}^k$ with non-zero Lebesgue measure,

$$\liminf_{n \to +\infty} \sup_{\theta \in K} \frac{D\left(P_\theta^n \mid Q_n\right)}{\log_2 n} \geqslant \frac{k}{2}.$$

A statistical consequence is given in the following corollary, whose proof is given at the end of this section.

Corollary 2.14 *Under the assumptions of the previous theorem, let $\widehat{P}_n$ be a sequence of estimators for the conditional law of X_n given $X_{1:n-1}$. Then, Lebesgue-almost everywhere in θ,*

$$\limsup_{n \to +\infty} n E_{P_\theta^{n-1}} D\left(P_\theta^n(. \mid X_{1:n-1}) \mid \widehat{P}_n(. \mid X_{1:n-1})\right) \geqslant \frac{k}{2}.$$

In particular, if $(X_i)_{i \in \mathbb{N}}$ are i.i.d., and if $\tilde{\theta}_n$ is a sequence of estimators of θ, Lebesgue-almost everywhere in θ,

$$\limsup_{n \to +\infty} n E_\theta D\left(P_\theta \mid P_{\tilde{\theta}_n}\right) \geqslant \frac{k}{2}.$$

Proof (Proof of Theorem 2.13). By Corollary 2.9, for all measurable subsets G of $\mathscr{X}^n$:

$$D(P_\theta^n \mid Q_n) \geqslant P_\theta^n(G) \log_2 \frac{P_\theta^n(G)}{Q_n(G)} + P_\theta^n(G^c) \log_2 \frac{P_\theta^n(G^c)}{Q_n(G^c)}.$$

But $P_\theta^n(G) \log_2 P_\theta^n(G) + P_\theta^n(G^c) \log_2 P_\theta^n(G^c) \geqslant -1$ and $Q_n(G^c) \leqslant 1$, entailing

$$D\left(P_\theta^n \mid Q_n\right) \geqslant P_\theta^n(G) \log_2 \frac{1}{Q_n(G)} - 1. \tag{2.7}$$

Let $a_n = \log_2 n / \sqrt{n}$, let K be a bounded subset of $\mathbb{R}^k$ and $G_{\theta,n}$ be the set

$$G_{\theta,n} = \left\{ x_{1:n} \ : \ \left\| \hat{\theta}_n (x_{1:n}) - \theta \right\| \leqslant a_n \right\}$$

where $\hat{\theta}_n (x_{1:n})$ is the estimator $\hat{\theta}_n$ computed with the observations $x_{1:n}$. By assumption,

$$P_{\theta}^n (G_{\theta,n}) = P_{\theta}^n \left\{ \sqrt{n} \| \hat{\theta}_n - \theta \| \leqslant \log_2 n \right\}$$

tends to 1 for λ-almost all θ, λ being the Lebesgue measure. For all $\varepsilon > 0$, let

$$B_n(\varepsilon, K) = \left\{ \theta \in K \ : \ P_{\theta}^n(G_{\theta,n}) < 1 - \varepsilon \right\}.$$

By dominated convergence,

$$\lim_{n \to +\infty} \lambda \big(B_n(\varepsilon, K) \big) = 0.$$

If $A_n(K)$ is the set

$$A_n(K) = \left\{ \theta \in K \ : \ Q_n(G_{\theta,n}) \geqslant a_n^k \log_2 n \right\},$$

we have $\lim_{n \to +\infty} \lambda (A_n(K)) = 0$. Indeed, if M_n is the maximum number of disjoint balls with radius a_n having their center inside $A_n(K)$, then $\lambda(A_n(K)) \leqslant V(k) a_n^k M_n$ where $V(k)$ is the volume of the ball of $\mathbb{R}^k$ with radius 2, and since the balls are disjoint, the $G_{\theta,n}$, where θ is the center of a ball, are disjoint too, so that:

$$1 \geqslant \sum_{\theta \text{ a center}} Q_n(G_{\theta,n}) \geqslant M_n a_n^k \log_2 n$$

and $\lambda(A_n(K)) \leqslant V(k) / \log_2 n$.

If now $\theta \in K \backslash (A_n(K) \cup B_n(\varepsilon, K))$, then by (2.7)

$$D \left(P_{\theta}^n \mid Q_n \right) \geqslant (1 - \varepsilon) \left(k \log_2 \frac{\sqrt{n}}{\log_2 n} - \log_2 \log_2 n \right) - 1$$

and thus for n large enough

$$D \left(P_{\theta}^n \mid Q_n \right) \geqslant (1 - 2\varepsilon) \frac{k}{2} \log_2 n.$$

It follows that for all bounded subset K in $\mathbb{R}^k$,

$$\left\{ \theta \in K \ : \ \limsup_{n \to +\infty} \frac{D \left(P_{\theta}^n \mid Q_n \right)}{\log_2 n} < \frac{k}{2} \right\}$$

has Lebesgue measure zero. We then take the union in an increasing sequence of subsets K covering $\mathbb{R}^k$.

Moreover, for all $\varepsilon > 0$ and for all K such that $\lambda(K) > 0$, for n large enough, the set $K \backslash (A_n(K) \cup B_n(\varepsilon, K))$ is not empty. Thus for n large enough

$$\sup_{\theta \in K} D\left(P_\theta^n \mid Q_n\right) \geqslant (1 - \varepsilon) \left(k \log_2 \frac{\sqrt{n}}{\log_2 n} - \log_2 \log_2 n\right) - 1$$

and

$$\liminf_{n \to +\infty} \sup_{\theta \in K} \frac{D\left(P_\theta^n \mid Q_n\right)}{\log_2 n} \geqslant \frac{k}{2}. \qquad \square$$

Proof (Proof of Corollary 2.14). Letting

$$Q_n(x_{1:n}) = \widehat{P}_1(x_1) \widehat{P}_2(x_2 \mid x_1) \ldots \widehat{P}_n(x_2 \mid x_{1:n-1}),$$

then

$$D\left(P_\theta^n \mid Q_n\right) = \sum_{\ell=1}^{n} E_{P_\theta^{\ell-1}} D\left(P_\theta^\ell(\cdot \mid X_{1:n-1}) \mid \widehat{P}_\ell(\cdot \mid X_{1:\ell-1})\right).$$

We argue by contradiction: if

$$\limsup n E_{P_\theta^{n-1}} D\left(P_\theta^n(\cdot \mid X_{1:n-1}) \mid \widehat{P}_n(\cdot \mid X_{1:n-1})\right) < \frac{k}{2},$$

then for some $\varepsilon > 0$ and for $\ell \geqslant \ell_0$,

$$E_{P_\theta^{\ell-1}} D\left(P_\theta^\ell(\cdot \mid X_{1:\ell-1}) \mid \widehat{P}_\ell(\cdot \mid X_{1:\ell-1})\right) \leqslant \left(\frac{k}{2} - \varepsilon\right) \frac{1}{\ell},$$

$$D\left(P_\theta^n \mid Q_n\right) \leqslant \sum_{\ell=1}^{\ell_0-1} E_{P_\theta^{\ell-1}} D\left(P_\theta^\ell(\cdot \mid X_{1:\ell n-1}) \mid \widehat{P}_\ell(\cdot \mid X_{1:\ell-1})\right) + \left(\frac{k}{2} - \varepsilon\right) \sum_{\ell=\ell_0}^{n} \frac{1}{\ell}$$

and

$$\limsup_{n \to +\infty} \frac{D\left(P_\theta^n \mid Q_n\right)}{\log_2 n} < \frac{k}{2}. \qquad \square$$

2.3.2 Bayesian Statistics, Jeffrey's Prior

The goal of this section is to understand (with heuristic arguments, references on precise results will be given at the end of the chapter) which *prior* distribution to choose on the parameter so as to obtain, by mixture, a "good" coding distribution. Such a *prior* distribution has to approach the one (if any) achieving the maximin, thus leading to a minimax coding distribution, by Theorem 2.12.

In the framework of the previous section, we endow Θ with a *prior* distribution with density $\nu(\theta)$ with respect to the Lebesgue measure. We assume that the statistical model is dominated, and denote by p_θ^n the density of P_θ^n with respect to the dominant measure. In Bayesian statistics, one considers the random variable $(\theta, X_{1:n})$ over $\Theta \times \mathscr{X}^n$ whose distribution has density $\nu(u)p_u^n(x_{1:n})$ in $(u, x_{1:n})$, and one is interested in the conditional distribution of θ given $X_{1:n}$, called the *posterior* distribution of θ, with density $\nu(\cdot \mid X_{1:n})$.

Let $P_\nu^n = \int P_\theta^n \nu(\theta)\mathrm{d}\theta$ be the distribution with density

$$p_\nu^n(x_{1:n}) = \int_\Theta p_\theta^n(x_{1:n})\nu(\theta)\mathrm{d}\theta.$$

The Bayesian redundancy is the supremum in ν of

$$\int_\Theta D\left(P_\theta^n \mid p_\nu^n\right)\nu(\theta)\mathrm{d}\theta = \int_\Theta E_\theta\left(\log_2 \frac{p_\theta^n(X_{1:n})}{p_\nu^n(X_{1:n})}\right)\nu(\theta)\mathrm{d}\theta$$

$$= \int_\Theta E_\theta\left(\log_2 \frac{\nu(\theta \mid X_{1:n})}{\nu(\theta)}\right)\nu(\theta)\mathrm{d}\theta.$$

Notions and results of asymptotic statistics mentioned below can be found, for instance, in van der Vaart [4]. We assume that the model is "regular", with Fisher information I_θ^n invertible.

The Bernstein-von Mises Theorem expresses the fact that, in good cases (in particular, for i.i.d. observations), the *posterior* distribution of θ is close to a Gaussian distribution centered at the maximum likelihood $\widehat{\theta}_{MV}$ and with variance the inverse of I_θ^n, namely $\frac{1}{n}I_\theta^{-1}$ in the i.i.d. case, I_θ being the Fisher information with one observation.

Here are some heuristic arguments to understand this result. In the i.i.d. case, denoting by $\ell_n(\theta)$ the log-likelihood, we have

$$\ell_n\left(\theta_0 + \frac{h}{\sqrt{n}}\right) = \ell_n(\theta_0) + h^T I_{\theta_0}\Delta_{n,\theta_0} - \frac{1}{2}h^T I_{\theta_0}h + o_{P_\theta^n}(1)$$

with

$$\Delta_{n,\theta_0} = I_\theta^{-1}\frac{1}{\sqrt{n}}\sum_{i=1}^n \dot{\ell}_{\theta_0}(X_i) = \sqrt{n}(\widehat{\theta}_{MV} - \theta_0) + o_{P_\theta^n}(1),$$

where $\dot{\ell}_{\theta_0}$ is the score function in θ_0. It follows that the *posterior* distribution of $H = \sqrt{n}(\theta - \theta_0)$ has density proportional to

$$\exp\left\{\ell_n\left(\theta_0 + \frac{h}{\sqrt{n}}\right) - \ell_n(\theta_0)\right\}\nu\left(\theta_0 + \frac{h}{\sqrt{n}}\right)$$

and thus is approximately proportional to

$$\exp\left\{h^T I_{\theta_0} \Delta_{n,\theta_0} - \frac{1}{2} h^T I_{\theta_0} h\right\},$$

so that its law is approximately $\mathcal{N}(\Delta_{n,\theta_0}, I_{\theta_0}^{-1})$, and the *posterior* distribution of θ is approximately $\mathcal{N}(\widehat{\theta}_{MV}, \frac{1}{n} I_{\theta_0}^{-1})$.

Then $\int_\Theta D\left(P_\theta^n \mid P_\nu^n\right) \nu(\theta) \mathrm{d}\theta$ is approximately equal to

$$\int_\Theta \left[\log_2 \frac{1}{\nu(\theta)} + E_\theta \log_2 d\mathcal{N}\left(\widehat{\theta}_{MV}; \frac{1}{n} I_\theta^{-1}\right)(\theta)\right] \nu(\theta) \mathrm{d}\theta$$

$$= \int_\Theta \left[E_\theta \log_2 \left(\frac{n^{k/2}\sqrt{\det(I_\theta)}}{(2\pi)^{k/2}\nu(\theta)} \exp -\frac{1}{2}(\theta - \widehat{\theta}_{MV})^T (nI_\theta)(\theta - \widehat{\theta}_{MV})\right)\right] \nu(\theta) \mathrm{d}\theta$$

$$= \frac{k}{2} \log_2 \frac{n}{2\pi} + \int_\Theta \left(\log_2 \frac{\sqrt{\det(I_\theta)}}{\nu(\theta)} - \frac{\log_2 e}{2} E_\theta[(\theta - \widehat{\theta}_{MV})^T (nI_\theta)(\theta - \widehat{\theta}_{MV})]\right) \nu(\theta) \mathrm{d}\theta$$

$$= \frac{k}{2} \log_2 \frac{n}{2\pi e} + \int_\Theta \left(\log_2 \frac{\sqrt{\det(I_\theta)}}{\nu(\theta)}\right) \nu(\theta) \mathrm{d}\theta$$

since $n(\theta - \widehat{\theta}_{MV})^T I_\theta (\theta - \widehat{\theta}_{MV})$ has approximately a chi-square distribution with k degrees of freedom under P_θ^n. Hence

$$\int_\Theta D\left(P_\theta^n \mid P_\nu^n\right) \nu(\theta) \mathrm{d}\theta \approx \frac{k}{2} \log_2 \frac{n}{2\pi e} + \log_2 \left[\int_\Theta \sqrt{\det(I_\theta)} \mathrm{d}\theta\right] - D\left(\nu \mid \nu_J\right),$$

where

$$\nu_J(\theta) = \frac{\sqrt{\det(I_\theta)}}{\int_\Theta \sqrt{\det(I_\theta)} \mathrm{d}\theta}$$

provided $\int_\Theta \sqrt{\det(I_\theta)} \mathrm{d}\theta$ is finite. Maximizing over ν, we get $\nu = \nu_J$, which is called Jeffrey's *prior* distribution.

Let us come back to memoryless sources over a finite alphabet of size k, with distribution $(\theta_1, \ldots, \theta_k)$ in $\mathscr{S}_k$ (the simplex of $\mathbb{R}^k$).

Setting $\theta = (\theta_1, \ldots, \theta_{k-1})$, with $\theta \in \mathbb{S}_k$, the log-likelihood is $\sum_{i=1}^k (\log_2 \theta_i) 1_{X=i}$ and the score function is

$$(\dot{\ell}_\theta)_i = \frac{1}{\theta_i} 1_{X=i} - \frac{1}{\theta_k} 1_{X=k}, \quad i = 1, \ldots, k-1.$$

The Fisher information is

$$I_\theta = \mathrm{Diag}\left(\frac{1}{\theta_i}\right)_{1 \leqslant i \leqslant k-1} + \frac{1}{\theta_k} \mathbf{1} \cdot \mathbf{1}^T,$$

where $\mathrm{Diag}(u_i)_{1 \leqslant i \leqslant k-1}$ is the diagonal matrix of size $(k-1) \times (k-1)$ with diagonal terms $u_1, \ldots, u_{k-1}$, and $\mathbf{1}$ is the column vector with 1 at all coordinates. The determinant of Fisher information is

$$\det (I_\theta) = \prod_{i=1}^{k} \frac{1}{\theta_i},$$

and consequently, the optimal *prior* distribution has density proportional to $\prod_{i=1}^{k} \frac{1}{\sqrt{\theta_i}}$ on $\mathscr{S}_k$. We will now be interested in this kind of distribution.

2.4 Dirichlet Mixtures

The *Dirichlet distribution* with parameter $\alpha = (\alpha_1, \ldots, \alpha_k)$ is the probability distribution over $\mathbb{S}_k$ with density

$$\frac{\Gamma(\alpha_1 + \cdots + \alpha_k)}{\Gamma(\alpha_1) \ldots \Gamma(\alpha_k)} \prod_{i=1}^{k} \theta_i^{\alpha_i - 1},$$

where $\theta_k = 1 - \sum_{i=1}^{k-1} \theta_i$ when $(\theta_1, \ldots, \theta_{k-1}) \in \mathbb{S}_k$. This is a probability density:

$$\Gamma(\alpha_1) \ldots \Gamma(\alpha_k) = \int_0^\infty x_1^{\alpha_1 - 1} e^{-x_1} dx_1 \ldots \int_0^\infty x_k^{\alpha_k - 1} e^{-x_k} dx_k$$

$$= \int_0^\infty du \int_{\mathbb{S}_k} d\theta_1 \ldots d\theta_{k-1} \prod_{i=1}^{k} \theta_i^{\alpha_i - 1} u^{\sum_{i=1}^{k} \alpha_i - k} e^{-u} u^{k-1}$$

by the change of variables $(x_1 \ldots, x_k) \mapsto (u, \theta_1, \ldots, \theta_{k-1})$ defined by $x_i = \theta_i u$ for all $i = 1, \ldots, k-1$ and $u = x_1 + \cdots + x_k$, with Jacobian u^{k-1}.

The Jeffrey's *prior* over $\mathbb{S}_k$ is thus the Dirichlet distribution ν with parameter $(\frac{1}{2}, \ldots, \frac{1}{2})$.

2.4.1 *Mixture Coding of Memoryless Sources*

We consider the class $\mathscr{C}$ of memoryless sources, i.e. sequences of i.i.d. random variables over $\mathscr{X}$. We now assume that $\mathscr{X}$ is finite of size k. Thus, $\mathscr{C}$ is the set of distributions $\mathbb{P}_\theta = (\theta_1, \ldots, \theta_k)^{\otimes \mathbb{N}}$, where $(\theta_1, \ldots, \theta_k) \in \mathscr{S}_k$ with $\theta = (\theta_1, \ldots, \theta_{k-1}) \in \mathbb{S}_k$.

We endow $\mathbb{S}_k$ with the Dirichlet distribution ν with parameter $(\frac{1}{2}, \ldots, \frac{1}{2})$ and define the $\mathbb{KT}$ distribution, called the Krichevsky–Trofimov distribution, over $\mathscr{X}^\mathbb{N}$ by Kolmogorov's Extension Theorem, with

$$\mathbb{KT}(x_{1:n}) = \int_{\mathbb{S}_k} \mathbb{P}_\theta(x_{1:n}) \, \nu(d\theta) \tag{2.8}$$

for all integers n.

Computation of this distribution is simple and recursive.

Proposition 2.15 *Letting, for all $i \in \mathcal{X}$, $n_i = \sum_{j=1}^n 1_{x_j=i}$, we have:*

$$\mathbb{KT}(x_{1:n}) = \frac{\Gamma(\frac{k}{2}) \prod_{i=1}^k \Gamma(n_i + \frac{1}{2})}{\Gamma(\frac{1}{2})^k \Gamma(n + \frac{k}{2})}, \quad \mathbb{KT}(a \mid x_{1:n}) = \frac{n_a(x_{1:n}) + \frac{1}{2}}{n + \frac{k}{2}}.$$

Note that recursive computation requires updating of the n_i (whose dependence on $x_{1:n}$ is omitted for ease of notation), which consists in adding 1 to n_i for $i = x_{n+1}$.

Proof (Proof of Proposition 2.15). We have

$$\mathbb{KT}(x_{1:n}) = \int_{\mathscr{S}_k} \prod_{i=1}^k \theta_i^{n_i} \, \nu(d\theta) = \int_{\mathscr{S}_k} \frac{\Gamma(\frac{k}{2})}{\Gamma(\frac{1}{2})^k} \prod_{i=1}^k \theta_i^{n_i - \frac{1}{2}} \, d\theta_1 \dots d\theta_{k-1}$$

and recognize the Dirichlet distribution with parameter $(n_1 + \frac{1}{2}, \dots, n_k + \frac{1}{2})$. Using $\Gamma(x+1) = x\Gamma(x)$, the formula for the conditional distribution follows. $\quad\blacksquare$

The following inequality (the Krichevsky-Trofimov Inequality for memoryless sources) is crucial.

Theorem 2.16 *For all $x_{1:n}$ in $\mathcal{X}^n$, we have*

$$0 \leqslant -\log_2 \mathbb{KT}(x_{1:n}) + \log_2 \widehat{P}_{x_{1:n}}(x_{1:n}) \leqslant \frac{k-1}{2} \log_2 n + 2,$$

where $\widehat{P}_{x_{1:n}}$ is the maximum likelihood over the class of memoryless sources.

Proof (Proof of Theorem 2.16). First,

$$\widehat{P}_{x_{1:n}}(x_{1:n}) = \prod_{i=1}^k \left(\frac{n_i}{n}\right)^{n_i}.$$

We start by showing that

$$\log_2 \widehat{P}(x_{1:n}) - \log_2 \mathbb{KT}(x_{1:n}) \leqslant \log_2 \left(\frac{\Gamma(n + \frac{k}{2})\Gamma(\frac{1}{2})}{\Gamma(n + \frac{1}{2})\Gamma(\frac{k}{2})}\right). \tag{2.9}$$

By Proposition 2.15,

$$\log_2 \left(\frac{\Gamma(n + \frac{k}{2})\Gamma(\frac{1}{2})}{\Gamma(n + \frac{1}{2})\Gamma(\frac{k}{2})} \right) = -\log_2 \mathrm{KT}(z_{1:n}),$$

where $z_{1:n}$ is the word formed by n copies of the same letter. We thus have to show that for all $x_{1:n}$, letting, for $i = 1, \ldots, k$, n_i be the number of occurrences of letter i in $x_{1:n}$:

$$\prod_{i=1}^{k} \left(\frac{n_i}{n} \right)^{n_i} \leqslant \frac{\mathrm{KT}(x_{1:n})}{\mathrm{KT}(z_{1:n})},$$

which can be rewritten as

$$\prod_{i=1}^{k} \left(\frac{n_i}{n} \right)^{n_i} \leqslant \frac{\prod_{i=1}^{k} \left[(n_i - \frac{1}{2})(n_i - \frac{3}{2}) \ldots \frac{1}{2} \right]}{(n - \frac{1}{2})(n - \frac{3}{2}) \ldots \frac{1}{2}} = \frac{\prod_{i=1}^{k} \left[2n_i(2n_i - 1) \ldots (n_i + 1) \right]}{2n(2n - 1) \ldots (n + 1)}.$$

To establish this inequality, it suffices to show that there exists a one-to-one map which associates to all $m = 1, \ldots, n$ a pair (i, j), $1 \leqslant i \leqslant k$, $1 \leqslant j \leqslant n_i$ such that

$$\frac{n_i}{n} \leqslant \frac{n_i + j}{n + m}. \qquad (2.10)$$

Now, (2.10) holds if and only if $j \geqslant n_i m / n$. So for m and i fixed, the number of j such that (2.10) holds is strictly larger than $n_i - n_i m / n$, and for m fixed, the number of pairs (i, j) such that (2.10) holds is strictly larger than

$$\sum_{i=1}^{k} \left(n_i - \frac{n_i m}{n} \right) = n - m.$$

We may thus associate to $m = n$ a pair (i, n_i), and recursively associate to $m = n - 1, n - 2, \ldots$ a pair (i, j) such that (2.10) holds and which has not already been associated, since, for a given m, the number of remaining possible pairs is always strictly larger than $n - m$, the number of already associated pairs, which concludes the proof of (2.9).

Then, using (2.4), we see that $\log_2 \left(\Gamma(n + \frac{k}{2})\Gamma(\frac{1}{2}) / \Gamma(n + \frac{1}{2})\Gamma(\frac{k}{2}) \right)$ is upper-bounded by

$$\left(n + \frac{k - 1}{2} \right) \log_2 \left(n + \frac{k}{2} \right) - n \log_2 \left(n + \frac{1}{2} \right) - \left(\frac{k - 1}{2} \right) \log_2 \frac{k}{2}$$

$$+ \frac{\log_2 e}{2} + \log_2 \sqrt{\pi} + \frac{\log_2 e}{12(n + \frac{k}{2})}.$$

But the second part of this quantity satisfies

$$\frac{\log_2 \mathrm{e}}{2} + \log_2 \sqrt{\pi} + \frac{\log_2 \mathrm{e}}{12(n + \frac{k}{2})} \leqslant 2,$$

and the first part is equal to

$$\frac{k-1}{2} \log_2 n + n \log_2 \frac{n + \frac{k}{2}}{n + \frac{1}{2}} - \frac{k-1}{2} \log_2 \mathrm{e} + \frac{k-1}{2} \log_2 \frac{(n + \frac{k}{2})\mathrm{e}}{\frac{1}{2}nk}.$$

Now

$$n \log_2 \frac{n + \frac{k}{2}}{n + \frac{1}{2}} = n \log_2 \left(1 + \frac{\frac{k-1}{2}}{n + \frac{1}{2}}\right) \leqslant \frac{k-1}{2} \log_2 \mathrm{e},$$

and

$$\frac{k-1}{2} \log_2 \frac{(n + \frac{k}{2})\mathrm{e}}{\frac{1}{2}nk} \leqslant 0$$

as soon as $(n + \frac{k}{2})\mathrm{e} \leqslant \frac{1}{2}nk$, that is $2/k + 1/n \leqslant 1/\mathrm{e}$, which holds for n and k greater than or equal to 9, in which case

$$\log_2 \left(\frac{\Gamma(n + \frac{k}{2})\Gamma(\frac{1}{2})}{\Gamma(n + \frac{1}{2})\Gamma(\frac{k}{2})}\right) \leqslant \frac{k-1}{2} \log_2 n + 2. \qquad (2.11)$$

For n and k smaller than 9, one directly checks (2.11). Combining inequalities (2.9) and (2.11) concludes the proof.

Remark 2.12 This entails that for all $\theta \in \mathbb{S}_k$ and all $x_{1:n}$,

$$\log_2 \mathbb{P}_\theta(x_{1:n}) - \log_2 \mathbb{KT}(x_{1:n}) \leqslant \frac{k-1}{2} \log_2 n + 2$$

so that if $\mathbb{P}_\theta^n$ is the distribution of $X_{1:n}$ under $\mathbb{P}_\theta$,

$$D(\mathbb{P}_\theta^n \,|\, \mathbb{KT}) \leqslant \frac{k-1}{2} \log_2 n + 2,$$

which shows that $\mathbb{KT}$ achieves the asymptotic bound of Rissanen's Theorem 2.13 for the class of memoryless sources.

2.4.2   *Mixture Coding of Context Tree Sources*

Stationary ergodic sources can be approximated by Markov chains of arbitrary order, this is what we used to prove the Shannon-Breiman-McMillan Theorem 1.10. If, for all integers m, $\mathcal{M}_m$ is the set of stationary Markovian sources of order m, distributions in $\mathcal{M}_m$ are parametrized by a parameter of dimension $|\mathcal{X}|^m(|\mathcal{X}| - 1)$ which grows exponentially fast in m. For more flexibility and sparsity, we introduce Markov chains with variable order, also called context tree sources, which we now describe.

The idea is to consider processes such that the occurrence probability of a letter given the entire past of the process only depends on a finite part of that past, and the length of this finite part itself depends on the past. In other words, if $\mathbb{P}$ is the distribution of the source $(X_n)_{n\in\mathbb{Z}}$,

$$\mathbb{P}(x_0 \mid x_{-\infty:-1}) = \mathbb{P}(x_0 \mid x_{-m:-1}) \,,$$

where m is a function of $x_{-\infty:-1}$.

Definition 2.13 A *complete suffix dictionary* $\mathcal{D}$ is a finite part of $\mathcal{X}^*$ such that for all sequence $x_{-\infty:-1}$, there exists a unique integer m such that $x_{-m:-1} \in \mathcal{D}$. The elements of $\mathcal{D}$ are then called *contexts*.

The *context function* is the function f which associates to an infinite sequence its context: $f(x_{-\infty:-1}) = x_{-m:-1} \in \mathcal{D}$.

A complete suffix dictionary is the set of leaves of a complete tree.

We denote by $\ell(\mathcal{D})$ the maximum length of a word of $\mathcal{D}$. This maximum length corresponds to the depth of the context tree representing $\mathcal{D}$.

Definition 2.14 A source $(X_n)_{n\in\mathbb{Z}}$ with distribution $\mathbb{P}$ is a *source with context tree* $\mathcal{D}$ if it is stationary and if for all $x_{-\infty:n}$,

$$\mathbb{P}\big(X_n = x_n \mid X_{-\infty:n-1} = x_{-\infty:n-1}\big) = \mathbb{P}\big(X_n = x_n \mid f(X_{-\infty:n-1}) = f(x_{-\infty:n-1})\big),$$

where f is the context function.

One says that, in the word $x_{-\infty:n-1}$, letter x_i occurs in context s, for some integer $i \leqslant n$ and some context $s \in \mathcal{D}$, if $x_{i-\ell(s):i-1} = s$.

One says that $(X_n)_{n\in\mathbb{Z}}$ is a *context tree source* if there exists a complete suffix dictionary $\mathcal{D}$ such that $(X_n)_{n\in\mathbb{Z}}$ is a source with context tree $\mathcal{D}$.

Note that a source with context tree $\mathcal{D}$ is Markovian of order $\ell(\mathcal{D})$.

Let $CT_{\mathcal{D}}$ be the set of sources with context tree $\mathcal{D}$. One may parametrize $CT_{\mathcal{D}}$ as follows. Let

$$\Theta_{\mathcal{D}} = \big\{(\theta^s)_{s\in\mathcal{D}} \; : \; \theta^s \in \mathbb{S}_{|\mathcal{X}|}\big\}.$$

The dimension of the parameter set $\Theta_{\mathcal{D}}$ is $|\mathcal{D}|(|\mathcal{X}| - 1)$.

We identify $\mathcal{X}$ with $\{1, \ldots, k\}$. If $\theta = (\theta^s)_{s\in\mathcal{D}}$, then $\theta^s = (\theta^s_i)_{1\leqslant i\leqslant k-1} \in \mathbb{S}_k$ for all $s \in \mathcal{D}$ and if $(X_n)_{n\in\mathbb{Z}}$ is a source with context tree $\mathcal{D}$ and distribution $\mathbb{P}_{\mathcal{D},\theta}$, we have, for all contexts $s \in \mathcal{D}$,

$$\mathbb{P}_{\mathcal{D},\theta}\big(X_0 = x \mid X_{-\ell(s):-1} = s\big) = \theta^s_x,$$

where as before $\theta^s_k = 1 - \sum_{i=1}^{k-1} \theta^s_i$.

One easily infers that

$$\mathbb{P}_{\mathscr{D},\theta}\big(X_{1:n}=x_{1:n}\,|\,X_{-\infty:0}=x_{-\infty:0}\big) = \prod_{i=1}^{n}\mathbb{P}_{\mathscr{D},\theta}\big(X_i=x_i\,|\,f(X_{-\infty:i-1})=f(x_{-\infty:i-1})\big)$$

$$= \prod_{s\in\mathscr{D}}\mathbb{P}_{\theta^s}\big(S^*(s,x_{1:n};x_{-\infty:0})\big),$$

where $S^*(s,x_{1:n};x_{-\infty:0})$ is the word obtained by concatenating the letters of $x_{1,n}$ occurring, in the word $x_{-\infty:n}$, in context s, and $\mathbb{P}_{\theta^s}$ is the distribution of the memoryless source with parameter θ^s. Note that $x_{-\infty:0}$ has to be introduced to determine contexts in which letters occur, but it is only used to determine the context in which the first letters of $x_{1:n}$ occur, namely at most for the x_i such that $i < \ell(\mathscr{D})$.

Let us define over $\Theta_{\mathscr{D}}$ the *prior*

$$\nu_{\mathscr{D}}\,(d\theta) = \otimes_{s\in\mathscr{D}}\,\nu\left(d\theta^s\right),$$

where ν is the Dirichlet distribution with parameter $(\frac{1}{2},\ldots,\frac{1}{2})$ over $\mathbb{S}_k$.

We then define the distribution $\mathbb{KT}_{\mathscr{D}}$ by Kolmogorov's Extension Theorem, with

$$\mathbb{KT}_{\mathscr{D}}\,(x_{1:n}\,|\,x_{-\infty:0}) = \int_{\Theta_{\mathscr{D}}}\mathbb{P}_{\mathscr{D},\theta}\,(x_{1:n}\,|\,x_{-\infty:0})\,\nu_{\mathscr{D}}\,(d\theta)$$

for all integers n. We thus have

$$\mathbb{KT}_{\mathscr{D}}\,(x_{1:n}\,|\,x_{-\infty:0}) = \prod_{s\in\mathscr{D}}\int_{\mathbb{S}_k}\mathbb{P}_{\theta^s}\big(S^*(s,x_{1:n};x_{-\infty:0})\big)\,\nu(d\theta^s)$$

and, using Proposition 2.15, we obtain the following proposition.

Proposition 2.17 *For all $x_{-\infty:n}$, we have:*

$$\mathbb{KT}_{\mathscr{D}}\,(x_{1:n}\,|\,x_{-\infty:0}) = \prod_{s\in\mathscr{D}}\mathbb{KT}\big(S^*(s,x_{1:n};x_{-\infty:0})\big)$$

with

$$\mathbb{KT}\big(S^*(s,x_{1:n};x_{-\infty:0})\big) = \frac{\Gamma(\frac{k}{2})}{\Gamma(\frac{1}{2})^k}\frac{\prod_{y\in\mathscr{X}}\Gamma(a_s^y(x_{1:n}\,|\,x_{-\infty:0})+\frac{1}{2})}{\Gamma(b_s(x_{1:n}\,|\,x_{-\infty:0})+\frac{1}{2})}$$

where:

- $a_s^y(x_{1:n}\,|\,x_{-\infty:0}) = \displaystyle\sum_{i=1}^{n}\mathbf{1}_{x_{i-\ell(s):i-1}=s,x_i=y}$ *is the number of occurrences of y in context s in word $x_{1:n}$;*
- $b_s(x_{1:n}\,|\,x_{-\infty:0}) = \displaystyle\sum_{y\in\mathscr{X}}a_s^y(x_{1:n}\,|\,x_{-\infty:0})$ *is the length of the word $S^*(s,x_{1:n};x_{-\infty:0})$.*

Thanks to these formulas, one may compute recursively $a_s^y(x_{1:n}|x_{-\infty:0})$ and $b_s(x_{1:n}|x_{-\infty:0})$, as well as $\mathbb{KT}(S^*(s, x_{1:n}; x_{-\infty:0}))$ when going from $x_{1:n}$ to $x_{1:n+1}$.

Note that $b_s(x_{1:n}|x_{-\infty:0}) = \ell(S^*(s, x_{1:n}; x_{-\infty:0}))$.

Let now γ be the real-valued function given, for all $x > 0$, by

$$\gamma(x) = \frac{k-1}{2}\log_2 x + 2.$$

Proposition 2.18 *For all $x_{-\infty:n}$, we have:*

$$-\log_2 \mathbb{KT}_{\mathscr{D}}(x_{1:n}|x_{-\infty:0}) \leqslant \inf_{\theta \in \Theta_{\mathscr{D}}}\left\{-\log_2 \mathbb{P}_{\mathscr{D},\theta}(x_{1:n}|x_{-\infty:0})\right\} + |\mathscr{D}|\gamma\left(\frac{n}{|\mathscr{D}|}\right).$$

Proof (Proof of Proposition 2.18). By the inequality for mixture coding of memoryless sources:

$$-\log_2 \mathbb{KT}\left(S^*(s, x_{1:n}; x_{-\infty:0})\right) \leqslant \inf_{\theta^s}\left\{-\log_2 \mathbb{P}_{\theta^s}\left(S^*(s, x_{1:n}; x_{-\infty:0})\right)\right\}$$
$$+\gamma\left(b_s(x_{1:n}|x_{-\infty:0})\right).$$

Hence

$$-\log_2 \mathbb{KT}_{\mathscr{D}}(x_{1:n}|x_{-\infty:0}) \leqslant \sum_{s \in \mathscr{D}}\inf_{\theta^s}\left\{-\log_2 \mathbb{P}_{\theta^s}\left(S^*(s, x_{1:n})\right)\right\} + \gamma\left(b_s(x_{1:n}|x_{-\infty:0})\right)$$

$$\leqslant \inf_{\theta \in \Theta_{\mathscr{D}}}\left\{-\log_2 \mathbb{P}_{\mathscr{D},\theta}(x_{1:n}|x_{-\infty:0})\right\} + \sum_{s \in \mathscr{D}}\gamma\left(b_s(x_{1:n}|x_{-\infty:0})\right)$$

$$\leqslant \inf_{\theta \in \Theta_{\mathscr{D}}}\left\{-\log_2 \mathbb{P}_{\mathscr{D},\theta}(x_{1:n}|x_{-\infty:0})\right\} + |\mathscr{D}|\gamma\left(\frac{\sum_{s \in \mathscr{D}}b_s(x_{1:n}|x_{-\infty:0})}{|\mathscr{D}|}\right),$$

by concavity of γ, and the fact that $\sum_{s \in \mathscr{D}}b_s(x_{1:n}|x_{-\infty:0}) = n$.

In order to use this coding procedure, one has to know the initial context $x_{-\infty:0}$. However, $x_{-\ell(\mathscr{D})+1:0}$ suffices to determine the words $S^*(s, x_{1:n}; x_{-\infty:0})$.

To code without the initial context, we encode with the uniform distribution all the x_i's not occurring in any context. This amounts to adding a context ε, in such a way that $S^*(\varepsilon, x_{1:n})$ is defined as the concatenation of the x_i's such that for all $0 \leqslant m < i-1$, $x_{m:i-1} \notin \mathscr{D}$. For $s \in \mathscr{D}$, $S^*(s, x_{1:n})$ is defined as the concatenation of the x_i's such that there exists $0 \leqslant m < i-1$ with $x_{m:i-1} = s$. Note that $S^*(s, x_{1:n})$ is not necessarily the same word as $S^*(s, x_{1:n}; x_{-\infty:0})$, since it does not contain the x_i's whose context is found thanks to some x_m's, $m < 0$.

Define

$$\mathbb{KT}_{\mathscr{D}}(x_{1:n}) = \mathbb{P}_U\left(S^*(\varepsilon, x_{1:n})\right)\prod_{s \in \mathscr{D}}\mathbb{KT}\left(S^*(s, x_{1:n})\right),$$

where $\mathbb{P}_U$ is the distribution of i.i.d. uniform random variables over $\mathscr{X}$, namely $\mathbb{P}_U = \mathbb{P}_{\theta_0}$ for $\theta_0 = (1/k, \ldots, 1/k)$.

One may now write

$$\mathbb{KT}_{\mathscr{D}}(x_{1:n}) = \left(\frac{1}{k}\right)^{\ell[S^*(\varepsilon, x_{1:n})]} \prod_{s \in \mathscr{D}} \mathbb{KT}\left(S^*(s, x_{1:n})\right), \qquad (2.12)$$

where

$$\mathbb{KT}\left(S^*(s, x_{1:n})\right) = \frac{\Gamma(\frac{k}{2})}{\Gamma(\frac{1}{2})^k} \frac{\prod_{y \in \mathscr{X}} \Gamma(a_s^y(x_{1:n}) + \frac{1}{2})}{\Gamma(b_s(x_{1:n}) + \frac{1}{2})},$$

$a_s^y(x_{1:n}) = \sum_{i=\ell(s)+1}^{n} \mathbf{1}_{x_{i-\ell(s):i-1}=s, x_i=y}$ and $b_s(x_{1:n} \mid x_{-\infty:0}) = \sum_{y \in \mathscr{X}} a_s^y(x_{1:n})$ are computed recursively.

The following proposition states the Krichevsky-Trofimov Inequality for context tree sources.

Proposition 2.19 *For all* $x_{-\infty:n}$,

$$-\log_2 \mathbb{KT}_{\mathscr{D}}(x_{1:n}) \leqslant \inf_{\theta \in \Theta_{\mathscr{D}}} \inf_{x_{-\ell(\mathscr{D})+1:0}} \left\{ -\log_2 \mathbb{P}_{\mathscr{D},\theta}\left(x_{1:n} \mid x_{-\ell(\mathscr{D})+1:0}\right) \right\}$$

$$+ |\mathscr{D}| \gamma\left(\frac{n}{|\mathscr{D}|}\right) + \ell(\mathscr{D}) \log_2 k.$$

Proof (Proof of Proposition 2.19). We have

$$-\log_2 \mathbb{KT}_{\mathscr{D}}(x_{1:n}) = -\log_2 \mathbb{P}_U\left(S^*(\varepsilon, x_{1:n})\right) + \sum_{s \in \mathscr{D}} -\log_2 \mathbb{KT}\left(S^*(s, x_{1:n})\right).$$

By Theorem 2.16,

$$\sum_{s \in \mathscr{D}} -\log_2 \mathbb{KT}\left(S^*(s, x_{1:n})\right) \leqslant \sum_{s \in \mathscr{D}} \left(\inf_{\theta^s}\left\{-\log_2 \mathbb{P}_{\theta^s}\left(S^*(s, x_{1:n})\right)\right\}\right) + \gamma\left(|S^*(s, x_{1:n})|\right)$$

$$\leqslant \inf_{\theta \in \Theta_{\mathscr{D}}} \inf_{x_{-\ell(\mathscr{D})+1:0}} \left\{ \sum_{s \in \mathscr{D}} -\log_2 \mathbb{P}_{\theta^s}\left(S^*(s, x_{1:n} \mid x_{-\ell(\mathscr{D})+1:0})\right) \right\}$$

$$+ |\mathscr{D}| \gamma\left(\frac{\sum_{s \in \mathscr{D}} |S^*(s, x_{1:n})|}{|\mathscr{D}|}\right),$$

by concavity of γ, and denoting by $S^*(s, x_{1:n} \mid x_{-\ell(\mathscr{D})+1:0})$ the concatenation of the x_i's, $i \geqslant 0$, such that there exists $-\ell(\mathscr{D}) + 1 \leqslant m < i - 1$, with $x_{m:i-1} \in \mathscr{D}$ (which was used previously). But $\sum_{s \in \mathscr{D}} |S^*(s, x_{1:n})| \leqslant n$ and γ is increasing, so

$$|\mathscr{D}| \gamma\left(\frac{\sum_{s \in \mathscr{D}} |S^*(s, x_{1:n})|}{|\mathscr{D}|}\right) \leqslant |\mathscr{D}| \gamma\left(\frac{n}{|\mathscr{D}|}\right).$$

On the other hand, if $m \geqslant \ell(\mathscr{D})$, x_m necessarily occurs in some context, which gives $|S^*(\varepsilon, x_{1:n})| \leqslant \ell(\mathscr{D})$ and

$$-\log_2 \mathbb{P}_U \left(S^*(\varepsilon, x_{1:n}) \right) = \left| S^*(\varepsilon, x_{1:n}) \right| \log_2 k \leqslant \ell(\mathscr{D}) \log_2 k. \qquad \qquad \square$$

Remark 2.15 It follows that for all $\theta \in \Theta_{\mathscr{D}}$ and all $x_{1:n}$,

$$-\log_2 \mathbb{KT}_{\mathscr{D}} (x_{1:n}) \leqslant -\log_2 \mathbb{P}_{\mathscr{D},\theta} (x_{1:n}) + |\mathscr{D}| \gamma \left(\frac{n}{|\mathscr{D}|} \right) + \ell(\mathscr{D}) \log_2 k$$

so that if $\mathbb{P}_{\mathscr{D},\theta}^n$ is the distribution of $X_{1:n}$ under $\mathbb{P}_{\mathscr{D},\theta}$,

$$D(\mathbb{P}_{\mathscr{D},\theta}^n \,|\, \mathbb{KT}_{\mathscr{D}}) \leqslant \frac{|\mathscr{D}|(k-1)}{2} \log_2 n - \frac{|\mathscr{D}|(k-1)}{2} \log_2 |\mathscr{D}| + 2|\mathscr{D}| + \ell(\mathscr{D}) \log_2 k,$$

entailing that $\mathbb{KT}_{\mathscr{D}}$ achieves the asymptotic bound in Rissanen's Theorem 2.13 for the class of sources with context tree $\mathscr{D}$.

2.4.3   Double Mixture and Universal Coding

Let π be some probability distribution over the set of all complete trees. The CTW distribution (Context Tree Weighting distribution) is defined by Kolmogorov's Extension Theorem, with

$$\text{CTW} (x_{1:n}) = \sum_{\mathscr{D}} \pi(\mathscr{D}) \mathbb{KT}_{\mathscr{D}} (x_{1:n}) \qquad \qquad (2.13)$$

for all integers n.

For $\alpha \leqslant 1/k$, one may define a probability distribution π_α over the set of all complete trees through a branching process: each node has k descendants with probability α, and zero descendants with probability $1 - \alpha$, so that

$$\pi_\alpha(\mathscr{D}) = \alpha^{|I|} (1 - \alpha)^{|\mathscr{D}|} ,$$

where I is the set of strict suffixes of words of $\mathscr{D}$. By induction over $|I|$,

$$|\mathscr{D}| = |I| (k - 1) + 1$$

so that

$$\pi_\alpha(\mathscr{D}) = \alpha^{\frac{|\mathscr{D}|-1}{k-1}} (1 - \alpha)^{|\mathscr{D}|} .$$

Denote by CTW_α the distribution defined by (2.13) with $\pi = \pi_\alpha$.

Since for all complete trees $\mathscr{D}$,

$$-\log_2 \text{CTW}_\alpha (x_{1:n}) \leqslant -\log_2 \pi_\alpha(\mathscr{D}) \mathbb{KT}_{\mathscr{D}} (x_{1:n}) ,$$

we obtain, thanks to Proposition 2.19, the following result.

Proposition 2.20 *For all integers n, for all $x_{1:n}$:*

$$-\log_2 \mathrm{CTW}_\alpha (x_{1:n}) \leqslant \inf_{\mathscr{D}} \left[\inf_{\theta \in \Theta_{\mathscr{D}}} \inf_{x_{-\ell(\mathscr{D})+1:0}} \left\{ - \log_2 \mathbb{P}_{\mathscr{D},\theta} \left(x_{1:n} \,|\, x_{-\ell(\mathscr{D})+1:0} \right) \right\} \right.$$

$$\left. + |\mathscr{D}| \gamma \left(\frac{n}{|\mathscr{D}|} \right) + \ell(\mathscr{D}) \log_2 k - \frac{|\mathscr{D}| - 1}{k-1} \log_2 \alpha - |\mathscr{D}| \log_2 (1 - \alpha) \right].$$

An important consequence of this result is weak universality of code CTW_α for the class of stationary ergodic sources.

Theorem 2.21 *If $(X_n)_{n \in \mathbb{N}}$ is a stationary ergodic source with distribution $\mathbb{P}$ and finite entropy rate, then $\mathbb{P}$-a.s.*

$$\lim_{n \to +\infty} -\frac{1}{n} \log_2 \mathrm{CTW}_\alpha (X_{1:n}) = H_* (\mathbb{P}).$$

Proof (Proof of Theorem 2.21). Let $\mathbb{P}^m$ be the m-Markovian approximation of $\mathbb{P}$. Taking $\mathscr{D}$ as the total tree with depth m, i.e. $\mathscr{D} = \mathscr{X}^m$, we have

$$-\log_2 \mathrm{CTW}_\alpha (X_{1:n}) \leqslant -\log_2 \mathbb{P}^m (X_{1:n}) + k^m \gamma \left(\frac{n}{k^m} \right) + m \log_2 k$$

$$- \frac{k^m - 1}{k - 1} \log_2 \alpha - k^m \log_2 (1 - \alpha),$$

hence $\mathbb{P}$ a.s.

$$\lim_{n \to +\infty} -\frac{1}{n} \log_2 \mathrm{CTW}_\alpha (X_{1:n}) \leqslant H^m,$$

where $H^m = E[-\log_2 \mathbb{P}^m (X_0 \,|\, X_{-m:-1})]$ and the proof is concluded by letting m tend to infinity.

An enjoyable property of CTW_α is that $\mathrm{CTW}_\alpha (x_{1:n})$ may be computed recursively (see [12]).

2.5 Renewal Sources

Rissanen's Theorem shows that, for memoryless sources and Markovian sources, the minimax redundancy is at least (in speed) of order $\log_2 n$ times half the number of parameters. By mixture coding, the speed over those classes is at most $\log_2 n$ times half the number of parameters.

On the other hand, the class of stationary ergodic sources has no weak speed.

Are there big enough classes (but not too much) with speed larger than $\log_2 n$? We will see examples when $\mathscr{X}$ is countably infinite. But this is possible even with $\mathscr{X} = \{0, 1\}$. I. Csiszár and P. Shields showed it for the class of renewal processes and A. Garivier showed that $\mathrm{CTW}_{1/2}$ is adaptive over this class.

Let us first introduce renewal sources with values in $\mathscr{X} = \{0, 1\}$. Let Q be a probability distribution over $\mathbb{N} \backslash 0$ with finite expectation $\mu = \sum_{t \geqslant 1} t Q(t)$. Let $\mathbb{P}_Q$ be the distribution (if it exists) of the stationary process $(X_n)_{n \geqslant 1}$ over $\{0, 1\}^{\mathbb{Z}}$ such that inter-arrivals of 1's are i.i.d. random variables with distribution Q. More precisely, if T_0 is the random arrival time of the first 1 and $T_0 + \cdots + T_i$, with $i \geqslant 1$, is the random arrival time of the $(i + 1)$st 1, namely

$$T_0 = \min \{ j \geqslant 1 \; : \; X_j = 1 \}, \quad T_{i+1} = \min \{ j \geqslant T_i + 1 \; : \; X_j = 1 \}, \quad i \geqslant 0,$$

then $(X_n)_{n \geqslant 1}$ is a stationary process and $(T_n)_{n \geqslant 1}$ is a sequence of i.i.d. random variables with distribution Q. Let us first show that such a process exists and that its distribution is well-defined. Let

$$R(t) = \sum_{u \geqslant t} Q(u).$$

If $x_{1:n} = 0^{t_0 - 1} 10^{t_1 - 1} 1 \ldots 10^{t_N - 1} 10^{t_{N+1} - 1}$ with $t_i \geqslant 1$ and $i = 0, \ldots, N$, then

$$\mathbb{P}_Q(x_{1:n}) = \mathbb{P}_Q(T_0 = t_0) \prod_{i=1}^{N} Q(t_i) R(t_{N+1}).$$

By stationarity, for $t \in \mathbb{N}^*$,

$$\mathbb{P}_Q(T_0 = t) = \mathbb{P}_Q(T_0 = t + 1) + \mathbb{P}_Q(T_0 = 1) Q(t),$$

which by induction gives

$$\mathbb{P}_Q(T_0 = t) = \mathbb{P}_Q(T_0 = 1) R(t),$$

and

$$\mathbb{P}_Q(T_0 = 1) = \frac{1}{\sum_{t \in \mathbb{N}^*} R(t)} = \frac{1}{\mu},$$

by Fubini's Theorem, so that for all $x_{1:n}$, $\mathbb{P}_Q(x_{1:n})$ is uniquely defined by

$$\mathbb{P}_Q(x_{1:n}) = \frac{1}{\mu} R(t_0) \prod_{i=1}^{N} Q(t_i) R(t_{N+1})$$

if $x_{1:n} = 0^{t_0 - 1} 10^{t_1 - 1} 1 \ldots 10^{t_N - 1} 10^{t_{N+1} - 1}$ with $t_i \geqslant 1$ and $i = 0, \ldots, N$. We now check that this defines a consistent sequence of probability distributions over $\mathscr{X}^n$ for $n \geqslant 1$, in order to define $\mathbb{P}_Q$ by Kolmogorov's Extension Theorem.

2.5.1 Redundancy of Renewal Processes

Let $\mathscr{R}$ be the class of stationary processes over $\{0, 1\}$ whose inter-arrival renewal distribution has finite expectation. I. Csiszár and P. Shields [13] showed the following.

Theorem 2.22 *There exist two positive constants c and C such that for all integers $n \geqslant 1$,*

$$c\sqrt{n} \leqslant \bar{R}_n(\mathscr{R}) \leqslant R_n^*(\mathscr{R}) \leqslant C\sqrt{n}.$$

Proof (Proof of Theorem 2.22). Let us first establish the upper bound. Let $\mathscr{Q}$ be the set of probability distributions over $\mathbb{N}^*$ with finite expectation. We have

$$R_n^*(\mathscr{R}) = \log_2\left(\sum_{x_{1:n}} \sup_{Q \in \mathscr{Q}} \mathbb{P}_Q(x_{1:n})\right).$$

Let n_j be the number of inter-arrival times equal to j in $x_{1:n}$:

$$n_j = \sum_{k=1:n-j} \mathbf{1}_{x_{k:k+j}=10...01}, \quad j = 1, \dots, n-1, \ n_j = 0, \ j \geqslant n.$$

Then for all $Q \in \mathscr{Q}$,

$$\mathbb{P}_Q(x_{1:n}) \leqslant \prod_{j=1}^{n-1} Q(j)^{n_j}.$$

Let $M = (n_1, \dots, n_{n-1})$ be the inter-arrivals vector, and let $\widehat{Q}_M$ be the element of $\mathscr{Q}$ maximizing $\prod_{j=1}^{n-1} Q(j)^{n_j}$:

$$\widehat{Q}_M(j) = \left(\frac{n_j}{\sum_j n_j}\right)^{n_j}, \quad j = 1, \dots, n-1, \ \widehat{Q}_M(j) = 0, \ j \geqslant n.$$

Let $A_n(t, M)$ be the set of $x_{1:n}$'s with inter-arrivals vector equal to M and with arrival time of the first 1 equal to t. Then

$$\sum_{x_{1:n} \in A_n(t,M)} \prod_{j=1}^{n-1} \widehat{Q}_M(j)^{n_j} = \mathbb{P}_{\widehat{Q}_M}\big((N_1, \dots, N_{n-1}) = M \mid T_0 = t\big) \leqslant 1,$$

where $(N_1, \dots, N_{n-1})$ is the inter-arrivals random vector:

$$N_j = \sum_{k=1:n-j} \mathbf{1}_{X_{k:k+j}=10...01}, \quad j = 1, \dots, n-1.$$

Hence

$$\sum_{x_{1:n}} \sup_{Q \in \mathscr{Q}} \mathbb{P}_Q(x_{1:n}) \leqslant \sum_{t=1}^{n} \sum_{M=(n_1,\ldots,n_{n-1})} \sum_{x_{1:n} \in A_n(t,M)} \prod_{j=1}^{n-1} \widehat{Q}_M(j)^{n_j},$$

and

$$R_n^*(\mathscr{R}) \leqslant \log_2 \left(n \times \text{number of possible } M \right).$$

The number of possible M is the number of $(n_1, \ldots, n_{n-1})$ such that $\sum_{j=1}^{n-1} j n_j \leqslant n - 1$, which is smaller than $\exp(C^{te}\sqrt{n-1})$ for some constant $C^{te} > 0$, by the Hardy-Ramanujan Theorem [14], so that for all integers $n \geqslant 1$,

$$R_n^*(\mathscr{R}) \leqslant \log_2 n + C^{te} \left(\log_2 e \right) \sqrt{n-1} \leqslant C\sqrt{n}$$

for some constant $C > 0$.

Moving on to the lower bound, let a_n be a strictly positive even integer and Θ be the set of subsets of $\{1, \ldots, a_n\}$ with size $\frac{1}{2}a_n$. If $\theta \in \Theta$, let $\mathbb{P}_\theta$ be the distribution of a renewal source whose inter-arrival distribution is the uniform distribution over θ.

Let now W be a uniform random variable over Θ and $(X_n)_{n \geqslant 1}$ be the process with conditional distribution $\mathbb{P}_\theta$ given $W = \theta$. By Theorem 2.12

$$\bar{R}_n(\mathscr{R}) \geqslant I(W; X_{1:n}).$$

But $I(W; X_{1:n}) = H(W) - H(W \mid X_{1:n})$, and, if $\binom{p}{q}$ is the number of subsets of size q of a set of size p:

$$H(W) = \log_2 |\Theta| = \log_2 \binom{a_n}{\frac{1}{2}a_n} = \log_2 \frac{\Gamma(a_n+1)}{\left[\Gamma(\frac{a_n}{2}+1)\right]^2} = a_n - \frac{1}{2}\log_2(a_n) + O(1)$$

using (2.4). Consequently,

$$\bar{R}_n(\mathscr{R}) \geqslant a_n - \frac{1}{2}\log_2(a_n) - H(W \mid X_{1:n}) - O(1).$$

Let now

$$\widehat{\theta}(x_{1:n}) = \{j : n_j > 0\}.$$

For all $x_{1:n} \in \mathscr{X}^n$, $H(W \mid X_{1:n} = x_{1:n})$ is smaller than the logarithm (in base 2) of the cardinality of the set of θ's in Θ such that $\mathbb{P}(W = \theta \mid X_{1:n} = x_{1:n}) > 0$, hence such that $\mathbb{P}(W = \theta, X_{1:n} = x_{1:n}) > 0$. But if θ and $x_{1:n}$ are such that $\mathbb{P}(W = \theta, X_{1:n} = x_{1:n})$ is positive, then $\widehat{\theta}(x_{1:n}) \subset \theta$. Hence

$$H(W \mid X_{1:n}) \leqslant E\left(\log_2 \left|\{\theta : \widehat{\theta}(X_{1:n}) \subset \theta\}\right|\right).$$

But

$$\left|\{\theta \ : \ \widehat{\theta}(x_{1:n}) \subset \theta\}\right| = \begin{pmatrix} a_n - |\widehat{\theta}(x_{1:n})| \\ \frac{1}{2}a_n - |\widehat{\theta}(x_{1:n})| \end{pmatrix}$$

and, for all integers a and p such that $a \geqslant p$,

$$\frac{1}{2^{a-p}}\begin{pmatrix} a - p \\ \frac{1}{2}a - p \end{pmatrix} = \frac{(a-p)(a-p-1)\ldots 2\cdot 1}{a(a-2)\ldots(a-2p+2)\cdot(a-2p)^2(a-2p-2)^2\ldots 2^2 1^2} \leqslant 1,$$

so that

$$\log_2\left(\frac{a_n - |\widehat{\theta}(x_{1:n})|}{\frac{1}{2}a_n - |\widehat{\theta}(x_{1:n})|}\right) \leqslant a_n - |\widehat{\theta}(x_{1:n})|,$$

and

$$I(W; X_{1:n}) \geqslant E\left(|\widehat{\theta}(X_{1:n})|\right) - \frac{1}{2}\log_2(a_n) - O(1).$$

Now,

$$E\left(|\widehat{\theta}(X_{1:n})|\right) = \frac{1}{|\Theta|}\sum_{\theta \in \Theta} E_\theta\left(|\widehat{\theta}(X_{1:n})|\right), \quad E_\theta\left(|\widehat{\theta}(X_{1:n})|\right) = \sum_{j \in \theta} \mathbb{P}_\theta\left(j \in \widehat{\theta}(X_{1:n})\right).$$

Since the inter-arrival distribution is supported by θ, we have $T_i \leqslant a_n$ for all integers $i \geqslant 0$. If κ_n is an integer smaller than $\frac{n}{a_n} - 1$, we have $\sum_{i=0}^{\kappa_n} T_i \leqslant n$, so that

$$\mathbb{P}_\theta\left(j \in \widehat{\theta}(X_{1:n})\right) \geqslant \mathbb{P}_\theta\left(\exists i \in \{1, \ldots, \kappa_n\}, \ T_i = j\right)$$

$$\geqslant \left[1 - \left(1 - \frac{2}{a_n}\right)^{\kappa_n}\right].$$

We obtain

$$\bar{R}_n(\mathscr{R}) \geqslant \frac{a_n}{2}\left[1 - \left(1 - \frac{2}{a_n}\right)^{\kappa_n}\right] - \frac{1}{2}\log_2(a_n) - O(1)$$

and the result follows by taking $a_n \sim \sqrt{n}$.

2.5.2 Adaptivity of CTW for Renewal Sources

Let

$$R_n^*(\mathrm{CTW}_{1/2}; \mathscr{R}) = \sup_{Q \in \mathscr{Q}} \sup_{x_{1:n}} -\log_2 \mathrm{CTW}_{1/2}(x_{1:n}) + \log_2 \mathbb{P}_Q(x_{1:n})$$

be the regret achieved over class $\mathscr{R}$ by the double mixture $\mathrm{CTW}_{1/2}$.

The following theorem shows that, up to a $\log_2 n$-factor, $\text{CTW}_{1/2}$ achieves the minimax over class $\mathscr{R}$. One says that $\text{CTW}_{1/2}$ is adaptive for the regret, up to a $\log_2 n$-factor, over class $\mathscr{R}$.

Theorem 2.23 (Garivier [15]) *There exist constants C_1 and C_2 such that for all $n \in \mathbb{N}$:*

$$C_1 \sqrt{n} \log_2 n \leqslant R_n^* \left(\text{CTW}_{1/2}; \mathscr{R} \right) \leqslant C_2 \sqrt{n} \log_2 n.$$

Proof (Proof of Theorem 2.23). *Upper bound.* For all context trees $\mathscr{D}$ and for all $x_{1:n}$,

$$-\log_2 \text{CTW}_{1/2} (x_{1:n}) \leqslant -\log_2 \mathbb{KT}_{\mathscr{D}} (x_{1:n}) - \log_2 \pi_{1/2} (\mathscr{D}).$$

Note that $\sup_{Q \in \mathscr{Q}} \log_2 \mathbb{P}_Q (x_{1:n})$ is equal to the supremum over the set of probabilities Q over $\{1, \ldots, n\}$, hence the supremum is attained. Let $\widehat{Q}_{x_{1:n}}$ be a probability attaining the supremum, so that

$$R_n^* \left(\text{CTW}_{1/2}; \mathscr{R} \right) = \sup_{x_{1:n}} -\log_2 \text{CTW}_{1/2} (x_{1:n}) + \log_2 \mathbb{P}_{\widehat{Q}_{x_{1:n}}} (x_{1:n}).$$

The proof consists in constructing a context tree $\mathscr{D}$, with a parameter that makes the distribution of the context tree source, at point $x_{1:n}$, close to $\mathbb{P}_{\widehat{Q}_{x_{1:n}}} (x_{1:n})$. This will make $-\log_2 \mathbb{KT}_{\mathscr{D}} (x_{1:n})$ close to $-\log_2 \mathbb{P}_{\widehat{Q}_{x_{1:n}}} (x_{1:n})$. In the remainder of the proof, $x_{1:n}$ will be fixed, and we will write $\mathbb{P}_{\widehat{Q}}$ for $\mathbb{P}_{\widehat{Q}_{x_{1:n}}}$. We also denote by $\widehat{R}$ the function defined by

$$\widehat{R}(t) = \sum_{u \geqslant t} \widehat{Q}(u), \quad t \in \mathbb{N}^*.$$

Let $\mathscr{D}$ be the context tree $\{0^k, 10^j, j = 0, \ldots, k-1\}$. Here, $|\mathscr{X}| = 2$, $\ell(\mathscr{D}) = k$, $|\mathscr{D}| = k+1$, thus

$$-\log_2 \text{CTW}_{1/2} (x_{1:n}) \leqslant -\log_2 \mathbb{KT}_{\mathscr{D}} (x_{1:n}) + 2k + 1.$$

We write

$$x_{1:n} = 0^{t_0-1} 1 \, 0^{t_1-1} 1 \, 0^{t_2-1} 1 \, \ldots \, 0^{t_N-1} 1 \, 0^{t_{N+1}-1}$$

and decompose $x_{1:n}$ into three words b, m and e, with $b = 0^{t_0-1}1$ if $t_0 \leqslant k$ and $b = 0^k$ otherwise (hence $|b| \leqslant k$), $e = 0^{t_{N+1}-1}$ and m such that $x_{1:n} = b \cdot m \cdot e$.

We now define the parameter θ of a context tree source $\mathbb{P}_{\mathscr{D}, \theta}$ as follows.

- For j such that $1 \leqslant j \leqslant k$,

$$\mathbb{P}_{\mathscr{D}, \theta}(1 \,|\, 10^{j-1}) = \frac{\widehat{Q}(j)}{\widehat{R}(j)}, \quad \mathbb{P}_{\mathscr{D}, \theta}(0 \,|\, 10^{j-1}) = \frac{\widehat{R}(j+1)}{\widehat{R}(j)}.$$

- $\mathbb{P}_{\mathscr{D},\theta}(1\,|\,0^k) = A/B$ where B is the number of symbols in $x_{1:n}$ occurring in context 0^k, i.e. $B = |S^*(0^k, x_{1:n})|$ and $A = \sum_{i=1}^{N+1} \mathbf{1}_{t_i > k}$ is the number of 1's among them. (In particular, $A \leqslant n/k$.)

Hence:

- for $1 \leqslant t \leqslant k$,

$$\mathbb{P}_{\mathscr{D},\theta}(0^{t-1}1\,|\,1) = \prod_{u=1}^{t-1} \mathbb{P}_{\mathscr{D},\theta}(0\,|\,10^{u-1})\,\mathbb{P}_{\mathscr{D},\theta}(1\,|\,10^{t-1})$$

$$= \prod_{u=1}^{t-1} \frac{\widehat{R}(u+1)}{\widehat{R}(u)}\,\frac{\widehat{Q}(t)}{\widehat{R}(t)} = \widehat{Q}(t) = \mathbb{P}_{\widehat{Q}}(0^{t-1}1\,|\,1);$$

- for $t > k$,

$$\mathbb{P}_{\mathscr{D},\theta}(0^{t-1}1\,|\,1) = \left(\prod_{u=1}^{k} \mathbb{P}_{\mathscr{D},\theta}(0\,|\,10^{u-1})\right)\mathbb{P}_{\mathscr{D},\theta}(0^{t-k}1\,|\,0^k)$$

$$= \widehat{R}(k+1)\,\mathbb{P}_{\mathscr{D},\theta}(0^{t-k}1\,|\,0^k)$$

$$\geqslant \widehat{Q}(t)\,\mathbb{P}_{\mathscr{D},\theta}(0^{t-k}1\,|\,0^k).$$

This yields, by Proposition 2.19,

$$\mathbb{P}_{\mathscr{D},\theta}(m \cdot e\,|\,b) \geqslant \prod_{i=1}^{N} \widehat{Q}(t_i)\,\widehat{R}(t_{N+1}) \prod_{i \geqslant 1:t_i > k} \mathbb{P}_{\mathscr{D},\theta}(0^{t_i-k}1\,|\,0^k)$$

$$\geqslant \mathbb{P}_{\widehat{Q}}(x_{1:n}) \prod_{i \geqslant 1:t_i > k} \mathbb{P}_{\mathscr{D},\theta}(0^{t_i-k}1\,|\,0^k).$$

Now

$$-\log_2 \mathrm{KT}_{\mathscr{D}}(x_{1:n}) = -\log_2 \mathbb{P}_U(b) - \log_2 \mathrm{KT}_{\mathscr{D}}(m \cdot e\,|\,b)$$

$$\leqslant |b| - \log_2 \mathbb{P}_{\mathscr{D},\theta}(m \cdot e\,|\,b) + |\mathscr{D}|\gamma\left(\frac{n}{|\mathscr{D}|}\right)$$

$$\leqslant k + (k+1)\gamma\left(\frac{n}{k+1}\right) - \log_2 \mathbb{P}_{\widehat{Q}}(x_{1:n})$$

$$- \log_2 \prod_{i \geqslant 1:t_i > k} \mathbb{P}_{\mathscr{D},\theta}(0^{t_i-k}1\,|\,0^k).$$

On the other hand,

$$\prod_{i \geqslant 1:t_i > k} \mathbb{P}_{\mathscr{D},\theta}(0^{t_i-k}1\,|\,0^k) = \left(\frac{A}{B}\right)^A \left(\frac{B-A}{B}\right)^{B-A}$$

so that

$$-\log_2 \prod_{i \geqslant 1 : t_i > k} \mathbb{P}_{\mathscr{D},\theta}(0^{t_i-k}1 \,|\, 0^k) = A \log_2 \left(\frac{B}{A}\right) + \left(B - A\right) \log_2 \frac{B}{B - A}.$$

But the function $x \mapsto (\frac{1}{x} - 1) \log_2(\frac{1}{1-x})$ is bounded over $[0, 1]$. So taking $x = A/B$, we get $(B - A) \log_2 \frac{B}{B-A} \leqslant D \cdot A$ for some constant $D > 0$.

Since $B \leqslant n$, we have $A \log_2(B/A) \leqslant A \log_2(n/A)$, and since $x \mapsto x \log_2 \frac{1}{x}$ is decreasing in a neighborhood of 0, taking $x = A/n \leqslant 1/k$ we get $A \log_2(n/A) \leqslant (n/k) \log_2 k$ for k large enough. Thus, for k large enough,

$$-\log_2 \prod_{i \geqslant 1 : t_i > k} \mathbb{P}_{\mathscr{D},\theta}(0^{t_i-k}1 \,|\, 0^k) \leqslant \frac{n}{k} \log_2 k + D\frac{n}{k}.$$

We obtain

$$-\log_2 \mathrm{CTW}_{1/2}(x_{1:n}) + \log_2 \mathbb{P}_{\widehat{Q}}(x_{1:n})$$
$$\leqslant 3k + 1 + \frac{n}{k} \log_2 k + D\frac{n}{k} + (k+1)\left[\frac{1}{2} \log_2 \left(\frac{n}{k+1}\right) + 2\right],$$

and taking $k = \lceil \sqrt{n} \rceil$, for some constant $C_2 > 0$,

$$-\log_2 \mathrm{CTW}_{1/2}(x_{1:n}) + \log_2 \mathbb{P}_{\widehat{Q}}(x_{1:n}) \leqslant C_2 \sqrt{n} \log_2 n.$$

Thus being true for all $x_{1:n}$,

$$R_n^*\left(\mathrm{CTW}_{1/2}; \mathscr{R}\right) \leqslant C_2 \sqrt{n} \log_2 n.$$

Lower bound. For all $x_{1:n}$

$$R_n^*\left(\mathrm{CTW}_{1/2}; \mathscr{R}\right) \geqslant -\log_2 \mathrm{CTW}_{1/2}(x_{1:n}) + \log_2 \mathbb{P}_{\widehat{Q}}(x_{1:n})$$
$$\geqslant \inf_{\mathscr{D}} -\log_2 \mathrm{KT}_{\mathscr{D}}(x_{1:n}) + \log_2 \mathbb{P}_{\widehat{Q}}(x_{1:n}).$$

We will choose a specific $x_{1:n}$ for which we can evaluate the right-hand side in this inequality.

Let $k = \lceil \sqrt{n} \rceil$, q be the quotient in the Euclidean division of $n - 1$ by $2k$ and ℓ be the remainder in that division. Then $\ell + 1 \leqslant 2k$ and $\frac{1}{2}(\sqrt{n} - 3) < q \leqslant \frac{1}{2}(\sqrt{n} + 1)$. Let

$$x_{1:n} = 0^\ell 1(0^{2k-1}1)^q.$$

Let us first note that taking Q to be the probability defined by $Q(2k) = 1$, we have $R(\ell) = 1$ and

$$\mathbb{P}_{\widehat{Q}}(x_{1:n}) \geqslant \mathbb{P}_Q(x_{1:n}) = \frac{1}{2k} \geqslant \frac{1}{2\sqrt{n}}.$$

Hence

$$-\log_2 \mathbb{KT}_{\mathscr{D}}(x_{1:n}) + \log_2 \mathbb{P}_{\widehat{Q}}(x_{1:n}) \geqslant -\sum_{s \in \mathscr{D}} \log_2 \mathbb{KT}\left(S^*(s, x_{1:n})\right) - \log_2\left(2\sqrt{n}\right)$$

by Eq. (2.12).

Let $\mathscr{D}$ be some context tree and let k_0 the largest integer m such that 0^m is in $\mathscr{D}$.

- If $k \leqslant k_0$, then $\mathscr{D}$ contains 0^{k-1} as an internal node, hence also $0^j 1$ for $j = 1, \ldots, k-1$ as nodes (whether internal or not). Letting S_j be the set of words of $\mathscr{D}$ stemming from $0^j 1$, we have

$$-\sum_{s \in \mathscr{D}} \log_2 \mathbb{KT}\left(S^*(s, x_{1:n})\right) \geqslant -\sum_{j=1}^{k-1} \sum_{s \in S_j} \log_2 \mathbb{KT}\left(S^*(0^j 1 s, x_{1:n})\right).$$

Note that the value of $\mathbb{KT}$ at a word does not depend on the letters' order and increases as letters are removed. Now concatenating $S^*(0^j 1 s, x_{1:n})$ for $s \in S_j$ with j fixed, we obtain at least 0^q (and possibly some 1's). On the other hand, for all integers q_i, $i = 1, \ldots, r$, we have

$$\mathbb{KT}(0^{q_1}) \ldots \mathbb{KT}(0^{q_r}) \leqslant \mathbb{KT}(0^{q_1 + \cdots + q_r}).$$

Hence

$$-\sum_{s \in S_j} \log_2 \mathbb{KT}\left(S^*(0^j 1 s, x_{1:n})\right) \geqslant -\log_2 \mathbb{KT}(0^q) \geqslant \frac{\log_2 e}{2} \log_2 q$$

since

$$\log_2 \mathbb{KT}(0^q) = \frac{\Gamma(1)\Gamma(q + \frac{1}{2})}{\Gamma(\frac{1}{2})^2 \Gamma(q + 1)},$$

and

$$\log_2 \mathbb{KT}(0^q) \leqslant \sum_{i=1}^{q} \log_2\left(\frac{2i-1}{2i}\right) \leqslant -(\log_2 e) \sum_{i=1}^{q} \frac{1}{2i} \leqslant -\frac{\log_2 e}{2} \log_2 q.$$

Consequently,

$$- \sum_{s \in \mathscr{D}} \log_2 \mathbb{KT}\big(S^*(s, x_{1:n})\big) \geq \frac{(k-1)\log_2 \mathrm{e}}{2} \log_2 q \geq \frac{(\sqrt{n}-2)\log_2 \mathrm{e}}{2} \log_2 \frac{\sqrt{n}-3}{2}.$$

- If $k_0 < k$, then $S^*(0^{k_0}, x_{1:n})$ contains at least $0^{kq}1^q$ up to a reordering of letters. Hence
$$- \log_2 \mathbb{KT}_{\mathscr{D}}(x_{1:n}) \geq - \log_2 \mathbb{KT}(0^{kq}1^q).$$

But $\mathbb{KT}$ is upper-bounded by the maximum likelihood of the memoryless sources' model, so

$$\mathbb{KT}(0^{kq}1^q) \leq \Big(\frac{kq}{kq+q}\Big)^{kq} \Big(\frac{q}{kq+q}\Big)^q$$

and we easily get

$$- \log_2 \mathbb{KT}_{\mathscr{D}}(x_{1:n}) \geq q \log_2(k+1) \geq \frac{\sqrt{n}-3}{4} \log_2 n.$$

In all cases, we have

$$- \log_2 \mathbb{KT}_{\mathscr{D}}(x_{1:n}) + \log_2 \mathbb{P}_{\widehat{Q}}(x_{1:n}) \geq \frac{\sqrt{n}-3}{2} \log_2 \frac{\sqrt{n}-3}{2}$$

and the lower bound follows.

2.6   Notes

Lempel-Ziv coding has several variants, and its properties have been intensely investigated. The first proof of universality was given by Ziv [16]. The proof given here entirely relies on the decomposition of words into pairwise distinct sequences. P. Shields' book [17] contains various references and more detailed analyses of this kind of decomposition (size and number of typical sequences in a decomposition into pairwise distinct sequences, return times, etc.), as well as a description of the link between context trees, entropy rate and its estimation.

Rissanen's Theorem and its variants led to numerous results establishing asymptotic lower bounds for the statistical risk. Jorma Rissanen was one of the first to introduce coding ideas and information theory in statistics. Among other things, he pioneered the MDL principle in model selection, which will be discussed in Chap. 4.

Evaluating maximin redundancy and identifying Jeffrey's prior as achieving the maximin was done by Clarke and Barron [18] for parametric regular models. This has a deep connection with Bayesian statistics, in which the reference prior is defined this way [19]. There now exist precise asymptotic developments for those quantities in finite alphabet settings, see for instance Szpankowski [20].

The Bernstein-von Mises Theorem in parametric frameworks is an old result. In non-parametric settings, it adds to the question of the speed of convergence that of the shape of the asymptotic distribution, see the recent results [21–24].

The mixture of discrete distributions through Dirichlet priors was denoted $\mathbb{KT}$ as a reference to Krichevsky and Trofimov [25]. The proof of Theorem 2.16 is due to Imre Csiszár [9].

One can find in Olivier Catoni's lectures [26] a detailed study of mixture distributions, properties and algorithms, references, and connections with statistical learning theory.

Context trees are popular in computer science and the idea of using variable order Markov chains in model selection for stationary ergodic processes originated from their analysis, see [27].

The CTW algorithm was invented by Willems et al. [12].

References

1. J. Ziv, A. Lempel, A universal algorithm for sequential data compression. IEEE Trans. Inform. Theory **23**, 337–343 (1977)
2. P. Shields, Universal redundancy rates don't exist. IEEE Trans. Inform. Theory **39**, 520–524 (1993)
3. Y.M. Shtarkov, Universal sequential coding of individual messages. Problemy Peredachi Informatsii **23**, 3–17 (1987)
4. A. van der Vaart, Asymptotic statistics, in *Cambridge Series in Statistical and Probabilistic Mathematics*, vol. 3. (Cambridge University Press, Cambridge, 1998). ISBN 0-521-49603-9; 0-521-78450-6
5. E. Whittaker, A. Watson, *A Course of Modern Analysis* (Cambridge University Press, Cambridge, 1996)
6. R. Dudley, *Real Analysis and Probability*, 2nd edn. (Cambridge University Press, New York, 2002)
7. M. Sion, On general minmax theorems. Pacific J. Math. **8**, 171–176 (1958)
8. D. Haussler, A general minimax result for relative entropy. IEEE Trans. Inform. Theory **43**, 1276–1280 (1997)
9. I. Csiszár. Class notes on information theory and statistics (University of Maryland, 1990)
10. J. Rissanen, Stochastic complexity and modeling. Ann. Statist. **14**, 1080–1100 (1986)
11. A. Barron, N. Hengartner, Information theory and superefficiency. Annals of Stat. **26**, 1800–1825 (1998)
12. F. Willems, Y. Shtarkov, T. Tjalkens, The context-tree weighting method: basic properties. IEEE Trans. Inform. Theory **41**, 653–664 (1995)
13. I. Csiszár, P.C. Shields, Redundancy rates for renewal and other processes. IEEE Trans. Inf. Theory **42**, 2005–2072 (1996)
14. G. Hardy, S. Ramanujan. Asymptotic formulæ in combinatory analysis (Proc. London Math. Soc. **17**(2) (1918), 75–115), in *Collected Papers of Srinivasa Ramanujan*, pp. 276–309 (AMS Chelsea Publ., Providence, RI, 2000)
15. A. Garivier, Redundancy of the context-tree weighting method on renewal and Markov renewal processes. IEEE Trans. Inform. Theory **52**, 5579–5586 (2006)
16. J. Ziv, Coding theorems for individual sequences. IEEE Trans. Inf. Theory **24**, 312–405 (1978)
17. P. Shields. The Ergodic theory of discrete sample paths, in *Graduate Studies in Mathematics*, vol. 13 (American Mathematical Society, 1996)

18. B. Clarke, A. Barron, Information-theoretic asymptotics of Bayes methods. IEEE Trans. Inf. Theory **36**, 453–471 (1990)
19. J. Bernardo, Reference posterior distributions for Bayesian inference. J. Roy. Statist. Soc. Ser. B **41**, 113–147 (1979)
20. W. Szpankowski, Average case analysis of algorithms on sequences, in *Wiley-Interscience Series in Discrete Mathematics and Optimization* (Wiley-Interscience, New York, 2001)
21. S. Boucheron, E. Gassiat, A Bernstein-von Mises theorem for discrete probability distributions. Electron. J. Stat. **3**, 114–148 (2009)
22. I. Castillo, A semiparametric Bernstein-von Mises theorem for Gaussian process priors. *Probab. Theory Related Fields*, **152**(1–2), 53–99 (2012). ISSN 0178-8051. http://dx.doi.org/10.1007/s00440-010-0316-5
23. V. Rivoirard, J. Rousseau, Bernstein-von Mises theorem for linear functionals of the density. Ann. Statist. **40**(3), 1489–1523 (2012). ISSN 0090-5364. http://dx.doi.org/10.1214/12-AOS1004
24. D. Bontemps, Bernstein-von Mises theorems for Gaussian regression with increasing number of regressors. Ann. Statist. **39**(5), 2557–2584 (2011). ISSN 0090-5364. http://dx.doi.org/10.1214/11-AOS912
25. R.E. Krichevsky, V.K. Trofimov, The performance of universal encoding. IEEE Trans. Inf. Theory **27**, 199–207 (1981)
26. O. Catoni, Statistical learning theory and stochastic optimization, in *Lecture Notes in Mathematics*. Lectures from the 31th Summer School on Probability Theory held in Saint-Flour, July 2001, vol. 1851 (Springer, Berlin, 2004)
27. P. Buhlmann, A.J. Wyner, Variable length Markov chains. Annals of Stat. **27**, 480–513 (1999)

Chapter 3
Universal Coding on Infinite Alphabets

Abstract When facing data from a "huge" alphabet, one may not be able to apply the previous results with satisfying theoretical guarantees, especially when those results are asymptotic. By a "huge alphabet", we mean for instance that within the data, some letters may not have occurred yet. To understand how to cope with such situations, we will be interested in the case where the alphabet is infinite. In a finite alphabet, we have seen that there exist universal codes over the class of stationary ergodic sources. For classes of memoryless or Markovian sources, minimax redundancy and regret are both asymptotically equivalent to half the number of parameters times the logarithm base 2 of the encoded word length. In the non-parametric class of renewal sources, minimax redundancy and regret have the same asymptotic speed, up to multiplicative constants. All of this does not extend to infinite alphabets: there is no weakly universal code over the class of stationary ergodic sources, and we will see examples of classes for which the regret is infinite whereas the minimax redundancy is not. The chapter starts with an encoding of the integers, which will be useful in the design of other codes. Thanks to a theorem due to John Kieffer, we show that there is no weakly universal code over the class of stationary ergodic sources with values in a countable alphabet. We then focus on memoryless sources (sequences of i.i.d. random variables) and make use of the Minimax-Maximin Theorem 2.12 to obtain lower bounds on the minimax redundancy of classes characterized by the decay of the probability measure at infinity. Another approach is to code in two steps: first, encode the observed alphabet (letters occurring in the data), then, encode what is known as the "pattern", containing information about the positions of letter repetitions, in their order of occurrence.

In this chapter, $\mathscr{X}$ is assumed to be countably infinite.

© Springer International Publishing AG, part of Springer Nature 2018   75
É. Gassiat, *Universal Coding and Order Identification by Model
Selection Methods*, Springer Monographs in Mathematics,
https://doi.org/10.1007/978-3-319-96262-7_3

3.1 Elias Coding of the Integers

Elias [1] proposed a prefix code of the integers, whose asymptotic code length is of order $\log_2 n$ for integer n. Note that binary writing of the integers does not give a prefix code. One may turn it into a prefix code by inserting a 0 between two symbols and a 1 at the end, but then the code length for integer n is of order $2 \log_2 n$.

Let us describe the code proposed by Elias. The code word for integer n, denoted $\mathscr{E}(n)$, is formed by concatenated subwords composed from right to left. First, the code word for 1 is $\mathscr{E}(1) = 0$. Then, if $n \geqslant 2$, we proceed as follows. The code word is terminated by the final subword 0. On the left, write n in binary, which gives a subword of length k. On the left of it, write $k - 1$ in binary, which gives a subword of length m, and so on until arriving at a subword of length 2. Concatenating those subwords yields $\mathscr{E}(n)$. Note that all subwords start with 1, except the last, which is just a 0.

To decode a codeword $\mathscr{E}(n)$, we proceed as follows. Read and decode subwords from left to right. The first subword is of size 2, decode it with binary decoding of the integers. Add 1 to the obtained integer: this gives the length of the next subword, which we decode (in binary), and so on, until the subword to decode start with a 0, this is the termination signal and the integer n is the last decoded integer. Since decoding allows us to identy the end of a codeword, this is a prefix code.

We obtain

$$\mathscr{E}(1) = 0, \quad \mathscr{E}(2) = 100, \quad \mathscr{E}(3) = 110, \quad \mathscr{E}(4) = 101000,$$
$$\mathscr{E}(5) = 101010, \quad \mathscr{E}(8) = 1110000$$

and so on.

If $L(n)$ is the length of codeword $\mathscr{E}(n)$, we have

$$L(n) = 1 + \sum_{m=1}^{k(n)} \left(\ell_m(n) + 1 \right),$$

where $\ell_1(j)$ is the integer part of $\log_2 j$, $\ell_{k+1}(j) = \ell_1(\ell_k(j))$ and $k(n)$ is the unique integer such that $\ell_{k(n)}(n) = 1$ for all $n \geqslant 2$.

For all integers p, if $2^p \leqslant n < 2^{p+1}$, $\ell_1(n) = p$, so the function $k(\cdot)$ is constant on the interval $[2^p; 2^{p+1}]$. Since it is a (weakly) increasing function satisfying, for all integers n,

$$k\left(2^n\right) = k(n) + 1,$$

we easily get $k(n) \leqslant \log_2 n$ for all integers $n \geqslant 2$. On the other hand, for all integers n,

$$L\left(2^n\right) = n + 1 + L(n).$$

Since $L(n) \leqslant 1 + k(n)[\log_2 n + 1]$, we have

$$\lim_{n \to +\infty} \frac{L(n)}{n} = 0,$$

which entails

$$\lim_{n \to +\infty} \frac{L(2^n)}{n} = 1.$$

It follows, using monotonicity of L, that:

$$\lim_{n \to +\infty} \frac{L(n)}{\log_2 n} = 1.$$

3.2 Universal Coding: Kieffer's Condition

Let Λ be a set of stationary ergodic sources' distributions over $\mathscr{X}$, each with finite entropy rate. Kieffer [2] proved the following theorem, which has an important consequence: there is no universal code over the class of stationary ergodic sources with values in a countable alphabet.

Theorem 3.1 *There exists a sequence* $(Q_n)_{n \geqslant 1}$ *of probabilities over* $\mathscr{X}^n$ *such that*

$$\forall \mathbb{P} \in \Lambda, \quad \lim_{n \to +\infty} \frac{1}{n} D(\mathbb{P}_n \mid Q_n) = 0$$

if and only if there exists a probability P^* *over* $\mathscr{X}$ *such that*

$$\forall \mathbb{P} \in \Lambda, \quad E_{\mathbb{P}^1}\left(-\log_2 P^*(X)\right) < +\infty.$$

Note that the condition is non-empty: there is no universal code over the class of stationary ergodic sources! Indeed, for all probabilities P^* over $\mathbb{N}$ with infinite support, there exists a probability P with finite entropy such that $E_P\left(-\log_2 P^*(X)\right) = +\infty$. It suffices to choose a strictly increasing sequence $(x_n)_{n \geqslant 1}$ such that for all n, $-\log_2 P^*(x_n) \geqslant 2^n$, and to choose P supported on the set $\{x_n, n \geqslant 1\}$, defined by $P(x_n) = \frac{1}{2^n}$.

It is remarkable that the necessary and sufficient condition given by the theorem only concerns the class of marginals of order 1 and not the dependency structure of sources.

Proof (Proof of necessity). Let $(Q_n)_{n \geqslant 1}$ be a sequence of probabilities over $\mathscr{X}^n$ such that

$$\forall \mathbb{P} \in \Lambda, \quad \lim_{n \to +\infty} \frac{1}{n} D(\mathbb{P}_n \mid Q_n) = 0.$$

Set $\mu_1 = Q_1$, and for all integers $k \geq 2$ and for all $x \in \mathscr{X}$

$$\mu_k(x) = \sup_{x_{2:k} \in \mathscr{X}^{k-1}} Q_k(xx_{2:k}).$$

Then, for all integers $k \geq 2$

$$\sum_{x \in \mathscr{X}} \mu_k(x) \leq \sum_{x \in \mathscr{X}} \sum_{x_{2:k} \in \mathscr{X}^{k-1}} Q_k(xx_{2:k}) = 1.$$

One may thus define, for all integers $k \geq 1$, the probability over $\mathscr{X}$ given by

$$\forall x \in \mathscr{X}, \quad R_k(x) = \frac{\mu_k(x)}{\sum_{y \in \mathscr{X}} \mu_k(y)},$$

and the probability over $\mathscr{X}$ given by

$$\forall x \in \mathscr{X}, \quad P^*(x) = \sum_{k \geq 1} \frac{1}{2^k} R_k(x).$$

For all integers $n \geq 1$ and all $x_{2:n} \in \mathscr{X}^{n-1}$,

$$-\log_2 Q_n(X_1 x_{2:n}) \geq -\log_2 \mu_n(X_1) \geq -\log_2 R_n(X_1) \geq -\log_2 P^*(X_1) - \log_2 2^n.$$

Let $\mathbb{P} \in \Lambda$. There exists an integer n_0 such that, as soon as $n \geq n_0$, the expectation $E_{\mathbb{P}}[-\log_2 Q_n(X_{1:n})]$ is finite. We then have

$$0 \leq E_{\mathbb{P}}[-\log_2 P^*(X_1)] \leq E_{\mathbb{P}}[-\log_2 Q_n(X_{1:n})] + \log_2 2^n < +\infty. \qquad \square$$

Proof (Proof of sufficiency). Let P^* be a probability over $\mathscr{X}$ such that

$$E_{\mathbb{P}^1}\left(-\log_2 P^*(X)\right) < +\infty$$

for all $\mathbb{P} \in \Lambda$.

Let us recall that a function s is the length function of a prefix code over $\mathscr{X}^n$ if and only if it is a function from $\mathscr{X}^n$ with values in $\mathbb{N}^*$ such that one may define a probability over $\mathscr{X}^n$ with mass $C2^{-s(x_{1:n})}$ for all $x_{1:n} \in \mathscr{X}^n$ and for some $C \geq 1$.

Let s^* be the length function of a prefix code over $\mathscr{X}$ such that for all $x \in \mathscr{X}$,

$$-\log_2 P^*(x) \leq s^*(x) < -\log_2 P^*(x) + 1.$$

For all integers n and all $\mathbb{P} \in \Lambda$, let $s_{n,\mathbb{P}}$ be the length function of a prefix code over $\mathscr{X}^n$ such that for all $x_{1:n} \in \mathscr{X}^n$,

$$- \log_2 \mathbb{P}_n \left(x_{1:n} \right) \leqslant s_{n,\mathbb{P}} \left(x_{1:n} \right) < - \log_2 \mathbb{P}_n \left(x_{1:n} \right) + 1.$$

For all integers n and for all $\mathbb{P} \in \Lambda$,

$$E_{\mathbb{P}} \left[- \log_2 (P^*)^{\otimes n} \left(X_{1:n} \right) \right] < + \infty,$$

hence there exists a finite subset $S_{\mathbb{P},n}$ of $\mathscr{X}^n$ such that

$$- \sum_{x_{1:n} \notin S_{\mathbb{P},n}} \mathbb{P}_n (x_{1:n}) \left[\sum_{i=1}^{n} \log_2 P^* (x_i) \right] \leqslant 1.$$

Let us now define

$$\sigma_{\mathbb{P},n} \left(x_{1:n} \right) = \begin{cases} \sum_{i=1}^{n} s^* (x_i) & \text{si } x_{1:n} \notin S_{\mathbb{P},n}, \\ s_{n,\mathbb{P}} \left(x_{1:n} \right) & \text{si } x_{1:n} \in S_{\mathbb{P},n}. \end{cases}$$

Then, $\sigma_{\mathbb{P},n}$ is the length function of a prefix code over $\mathscr{X}^n$ satisfying

$$\frac{1}{n} E_{\mathbb{P}} \left[\sigma_{\mathbb{P},n} \left(X_{1:n} \right) \right] \leqslant \frac{1}{n} H \left(X_{1:n} \right) + \frac{2}{n}.$$

Let $R_{\mathbb{P},n}$ be the corresponding probability over $\mathscr{X}^n$, which is such that:

$$\frac{1}{n} E_{\mathbb{P}} \left[- \log_2 R_{\mathbb{P},n} \left(X_{1:n} \right) \right] \leqslant \frac{1}{n} H \left(X_{1:n} \right) + \frac{2}{n}.$$

Now, the set $\{ R_{\mathbb{P},n}, \mathbb{P} \in \Lambda, n \in \mathbb{N} \}$ is countable. Indeed, let n be fixed. If S is a finite subset of $\mathscr{X}^n$, the set $\mathscr{L}_{S,n}$ of possible code length functions over S is countable. The union over finite subsets S of the $\mathscr{L}_{S,n}$ is countable, and the union over n of those sets is also countable.

We list this set as $\{ R^{(i)}, i \in \mathbb{N} \}$, in such a way that $R^{(i)}$ is a coding distribution over $\mathscr{X}^{n_i}$, n_i tends to infinity with i, and for all $\mathbb{P} \in \Lambda$, there exists an infinity of i's for which

$$\frac{1}{n_i} E_{\mathbb{P}} \left[- \log_2 R^{(i)} \left(X_{1:n_i} \right) \right] \leqslant \frac{1}{n_i} H \left(X_{1:n_i} \right) + \frac{2}{n_i}. \tag{3.1}$$

We then define for all i a sequence $(Q_k^{(i)})_{k \in \mathbb{N}}$ of coding distributions over $\mathscr{X}^k$ such that, if m_k is the remainder in the Euclidean division of k by n_i,

$$Q_k^{(i)} = R^{(i)} \otimes R^{(i)} \otimes \cdots \otimes R^{(i)} \otimes P^{* \otimes m_k}.$$

For all $\mathbb{P} \in \Lambda$, we have

$$\frac{1}{k} E_{\mathbb{P}} \left[- \log_2 Q_k^{(i)} (X_{1:k}) \right] \leqslant \frac{1}{n_i} E_{\mathbb{P}} \left[- \log_2 R^{(i)} \left(X_{1:n_i} \right) \right] + \frac{n_i}{k} E_{\mathbb{P}} \left[- \log_2 P^* (X_1) \right].$$

Finally, for integer n, denoting by J_n the integer such that $2^{J_n-1} < n \leqslant 2^{J_n}$, we may define, with $c_n \geqslant 1$, the coding distribution Q_n over $\mathcal{X}^n$ by

$$Q_n(x_{1:n}) = c_n 2^{-J_n} \max_{1 \leqslant i \leqslant n} Q_n^{(i)}(x_{1:n}).$$

For all $\mathbb{P} \in \Lambda$, clearly

$$\frac{1}{n} E_{\mathbb{P}}\left[-\log_2 Q_n(X_{1:n})\right] \geqslant \frac{1}{n} H(X_{1:n})$$

but also

$$\frac{1}{n} E_{\mathbb{P}}\left[-\log_2 Q_n(X_{1:n})\right] \leqslant \frac{J_n}{n} + \frac{1}{n} \inf_{1 \leqslant i \leqslant n} E_{\mathbb{P}}\left[-\log_2 Q_n^{(i)}(X_{1:n})\right]$$

and thus

$$\frac{1}{n} E_{\mathbb{P}}\left[-\log_2 Q_n(X_{1:n})\right]$$
$$\leqslant \inf_{1 \leqslant i \leqslant n}\left\{\frac{1}{n_i} E_{\mathbb{P}}\left[-\log_2 R^{(i)}(X_{1:n_i})\right] + \frac{n_i}{n} E_{\mathbb{P}}\left[-\log_2 P^*(X_1)\right]\right\} + \frac{J_n}{n}$$
$$\leqslant \inf_{i \in I \cap \{1,\ldots,n\}}\left\{\frac{1}{n_i} H(X_{1:n_i}) + \frac{2}{n_i} + \frac{n_i}{n} E_{\mathbb{P}}\left[-\log_2 P^*(X_1)\right]\right\} \frac{J_n}{n},$$

where I is the (infinite) set of i's such that (3.1) holds, so that

$$\lim_{n \to +\infty} \frac{1}{n} E_{\mathbb{P}}\left[-\log_2 Q_n(X_{1:n})\right] = H_*(\mathbb{P}). \qquad \square$$

3.3 Generalities on Redundancies and Regrets

Let Λ be a set of distributions of stationary ergodic sources with finite entropy. Let us recall that for all integers n, $R_n^*(\Lambda)$ denotes the regret over the class of n-dimensional marginals of sources of Λ (see Definition 2.6) and $\overline{R}_n(\Lambda)$ denotes the minimax redundancy over the class of n-dimensional marginals of sources of Λ (see Definition 2.4).

Proposition 3.2 *The sequences $(R_n^*(\Lambda))_{n \geqslant 1}$ and $(\overline{R}_n(\Lambda))_{n \geqslant 1}$ are (weakly) increasing. If furthermore Λ is a set of distributions of memoryless sources, then $(R_n^*(\Lambda))_{n \geqslant 1}$ and $(\overline{R}_n(\Lambda))_{n \geqslant 1}$ are sub-additive sequences, so that*

$$\lim_{n \to +\infty} \frac{R_n^*(\Lambda)}{n} = \inf_{n \in \mathbb{N}} \frac{R_n^*(\Lambda)}{n} \leqslant R_1^*(\Lambda),$$

$$\lim_{n \to +\infty} \frac{\overline{R}_n(\Lambda)}{n} = \inf_{n \in \mathbb{N}} \frac{\overline{R}_n(\Lambda)}{n} \leqslant \overline{R}_1(\Lambda),$$

and we have

$$R_n^*(\Lambda) < \infty \iff \sum_{x \in \mathcal{X}} \sup_{\mathbb{P} \in \Lambda} \mathbb{P}_1(x) < +\infty$$

Proof (Proof of Proposition 3.2). Let W be a random variable with values in Λ and $(X_n)_{n \geqslant 1}$ be a source with law W conditionally on W. For all integers $n \geqslant 1$,

$$I(W; X_{1:n+1}) = H(W) - H(W \mid X_{1:n+1}) \geqslant H(W) - H(W \mid X_{1:n}) = I(W; X_{1:n}).$$

Taking the supremum in W on each side of the inequality, we obtain, thanks to Theorem 2.12:

$$\overline{R}_{n+1}(\Lambda) \geqslant \overline{R}_n(\Lambda).$$

As for the regret, for all $x_{1:n} \in \mathcal{X}^n$,

$$\mathbb{P}_n(x_{1:n}) = \sum_{y \in \mathcal{X}} \mathbb{P}_{n+1}(x_{1:n}y),$$

hence $\sup_{\mathbb{P} \in \Lambda} \mathbb{P}_n(x_{1:n}) \leqslant \sum_{y \in \mathcal{X}} \sup_{\mathbb{P} \in \Lambda} \mathbb{P}_{n+1}(x_{1:n}y)$ and

$$\sum_{x_{1:n} \in \mathcal{X}^n} \sup_{\mathbb{P} \in \Lambda} \mathbb{P}_n(x_{1:n}) \leqslant \sum_{x_{1:n} \in \mathcal{X}^n} \sum_{y \in \mathcal{X}} \sup_{\mathbb{P} \in \Lambda} \mathbb{P}_{n+1}(x_{1:n}y) = \sum_{x_{1:n+1} \in \mathcal{X}^{n+1}} \sup_{\mathbb{P} \in \Lambda} \mathbb{P}_{n+1}(x_{1:n+1}),$$

so that $R_{n+1}^*(\Lambda) \geqslant R_n^*(\Lambda)$.

Let us now assume that Λ is a set of distributions of memoryless sources.

Let n, m be two integers and W as above. Since sources of Λ are memoryless, $X_{1:n}$ and $X_{n+1:n+m}$ are independent conditionally on W and

$$\begin{aligned} I(X_{n+1:n+m}; W \mid X_{1:n}) &= H(X_{n+1:n+m} \mid X_{1:n}) - H(X_{n+1:n+m} \mid X_{1:n}, W) \\ &= H(X_{n+1:n+m} \mid X_{1:n}) - H(X_{n+1:n+m} \mid W) \\ &\leqslant H(X_{n+1:n+m}) - H(X_{n+1:n+m} \mid W) \\ &= I(X_{n+1:n+m}; W). \end{aligned}$$

Then, by stationarity,

$$\begin{aligned} I(X_{1:n+m}; W) &= I(X_{1:n}; W) + I(X_{n+1:n+m}; W \mid X_{1:n}) \\ &\leqslant I(X_{1:n}; W) + I(X_{n+1:n+m}; W) = I(X_{1:n}; W) + I(X_{1:m}; W). \end{aligned}$$

Taking the supremum in W on each side of the inequality establishes sub-additivity of the sequence of minimax redundancies.

As for the regret, for all $x_{1:n} \in \mathcal{X}^n$, for all $y_{1:m} \in \mathcal{X}^m$ and for all $\mathbb{P} \in \Lambda$, we have

$$\mathbb{P}_{n+m} (x_{1:n} y_{1:m}) = \mathbb{P}_n (x_{1:n}) \mathbb{P}_m (y_{1:m}),$$

hence

$$\sup_{\mathbb{P} \in \Lambda} \mathbb{P}_{n+m} (x_{1:n} y_{1:m}) \leqslant \sup_{\mathbb{P} \in \Lambda} \mathbb{P}_n (x_{1:n}) \sup_{\mathbb{P} \in \Lambda} \mathbb{P}_m (y_{1:m}).$$

Sub-additivity of the sequence of regrets follows by summing over $x_{1:n}$ and $y_{1:m}$ and taking the logarithm in this inequality.

The asymptotic results then follow from Lemma 1.5.

When $\mathcal{X}$ is finite, we only encountered classes Λ for which minimax redundancy and regret had the same asymptotic equivalent. When $\mathcal{X}$ is infinite, one easily encounters examples where this is not the case. From now on, we assume

$$\mathcal{X} = \mathbb{N}.$$

Proposition 3.3 *Let f be a non-negative function, strictly decreasing over $\mathbb{N}$ and such that $f(1) < 1$. For all $k \in \mathbb{N}$, let p_k be the probability over $\mathbb{N}$ defined by:*

$$p_k(\ell) = \begin{cases} 1 - f(k) & \text{if } \ell = 0; \\ f(k) & \text{if } \ell = k; \\ 0 & \text{otherwise.} \end{cases}$$

Let $\Lambda^1 = \{p_1, p_2, \ldots\}$ and Λ be the set of memoryless sources with order-1 marginal in Λ^1. Then

$$\lim_{k \to +\infty} f(k) \log_2 k = +\infty \iff \bar{R}_n (\Lambda) = +\infty.$$

Remark 3.1 If, for all integers $k \geqslant 3$, we take $f(k) = 1/\log_2 k$, then $\bar{R}_n (\Lambda)$ is finite for all n. But since $\sum_{k \geqslant 0} f(k) = +\infty$, it follows that $R_n^* (\Lambda)$ is infinite for all n by Proposition 3.2.

Proof (Proof of Proposition 3.3). Assume that $\lim_{k \to +\infty} f(k) \log_2 k = +\infty$. Let m be some integer and W be a uniform random variable over $\{1, 2, \ldots, m\}$. Let X be a random variable with values in $\mathbb{N}$ and distribution p_k conditionally on $W = k$, and let $Z = \mathbf{1}_{X=W}$. Then $H(Z \mid W, X) = 0$, hence

$$H(W, Z \mid X) = H(W \mid X) + H(Z \mid W, X) = H(W \mid X).$$

On the other hand, $H(W \mid X, Z = 1) = 0$, so

$$\begin{aligned} H(W \mid X) &= H(Z \mid X) + H(W \mid Z, X) \\ &\leqslant 1 + \mathbb{P}(Z = 0) H(W \mid X, Z = 0) + \mathbb{P}(Z = 1) H(W \mid X, Z = 1) \\ &\leqslant 1 + (1 - f(m)) \log_2 m. \end{aligned}$$

Hence,

$$\bar{R}(\Lambda^1) \geqslant I(W, X) \geqslant \log_2 m - (1 - f(m)) \log_2 m - 1 = f(m) \log_2 m - 1$$

and $\bar{R}_1(\Lambda) = +\infty$.

Assume now that there exists a constant $C > 0$ such that $f(k) \log_2 k \leqslant C$ for all integers k. Let us show that $\bar{R}_1(\Lambda) < +\infty$ (which is sufficient since Λ consists of distributions of memoryless sources). Let Q be the probability over $\mathbb{N}$ defined by $Q(0) > 0$, $Q(1) > 0$ and $Q(k) = A/((1 \vee (k(\log_2 k)^2))$ for $k \geqslant 2$, where A is the normalizing constant. Then for all $k \geqslant 3$

$$D(p_k \mid Q) = (1 - f(k)) \log_2 \frac{(1 - f(k))}{Q(0)} + f(k) \log_2 \left(\frac{f(k)k(\log_2 k)^2}{A} \right)$$
$$\leqslant -\log_2 Q(0) + C + f(k) \left(2 \log_2^{(2)}(k) - \log_2(A) \right)$$
$$\leqslant C + \log_2 \frac{C^2}{A Q(0)}.$$

Hence

$$\bar{R}(\Lambda^1) \leqslant \left(C + \log_2 \frac{C^2}{A Q(0)} \right) \vee D(p_1, Q) \vee D(p_2, Q) < \infty. \qquad \square$$

3.4 Envelop Classes of Memoryless Sources

Let f be a function over $\mathbb{N}$ with values in $[0, 1]$. The envelop class Λ_f defined by f is the set of distributions of memoryless sources with order-1 marginal dominated by f:

$$\Lambda_f = \{ \mathbb{P} = P^{\otimes \mathbb{N}} : \forall x \in \mathbb{N}, \ P(x) \leqslant f(x) \}.$$

3.4.1 Generalities

Theorem 3.4 *We have the following equivalences:*

$$\bar{R}_n (\Lambda_f) < \infty \iff R_n^* (\Lambda_f) < \infty \iff \sum_{k \in \mathbb{N}} f(k) < \infty.$$

Proof (Proof of Theorem 3.4). By Proposition 3.2, it only remains to show that

$$\sum_{k \in \mathbb{N}} f(k) = \infty \implies \bar{R}_1 (\Lambda_f) = \infty.$$

Let $(h_i)_{i\in\mathbb{N}}$ be the sequence of integers defined recursively by $h_0 = 0$ and

$$h_{i+1} = \min\left\{h \in \mathbb{N}: \sum_{k=h_i+1}^{h} f(k) > 1\right\}.$$

Let P_i be the probability over $\mathbb{N}$ with support $\{h_i + 1, \ldots, h_{i+1}\}$ given by

$$P_i(m) = \frac{f(m)}{\sum_{k=h_i+1}^{h_{i+1}} f(k)} \quad \text{for } m \in \{h_i + 1, \ldots, h_{i+1}\}.$$

Then $P_i^{\otimes\mathbb{N}}$ belongs to Λ_f. Let W be a random variable with values in $\mathbb{N}$ and X a random variable with values in $\mathbb{N}$ and distribution P_W conditionally on W. Then $H(W \mid X) = 0$ and taking W such that $H(W) = \infty$, we get

$$\overline{R}_1\left(\Lambda_f\right) = \infty.$$ □

Theorem 3.5 *Let Λ be a set of memoryless sources and $\overline{F}_\Lambda$ defined by*

$$\overline{F}_\Lambda(u) = \sum_{k>u} \sup_{\mathbb{P}\in\Lambda} \mathbb{P}^1(k).$$

Then

$$R_n^*(\Lambda) \leqslant \inf_{u:u\leqslant n}\left[n\overline{F}_\Lambda(u)\log_2 e + \frac{1}{2}(u-1)\log_2 n\right] + 2$$

and

$$R_n^*(\Lambda) < \infty \iff R_n^*(\Lambda) = o(n).$$

Proof (Proof of Theorem 3.5). The last statement follows from the first by taking $(u_n)_{n\geqslant 1}$ such that $u_n \to +\infty$ and $(u_n \log_2 n)/n \to 0$ when $n \to +\infty$.

Let u be an integer and $x \in \mathbb{N}^n$. We decompose x into two non-contiguous sequences z, containing the letters of x that are strictly larger than u, and y, containing the letters of x that are smaller than or equal to u. Now, for memoryless sources, $\sup_{\mathbb{P}\in\Lambda} \mathbb{P}_n(x)$ does not depend on the order of symbols in x, so

$$\sum_{x\in\mathbb{N}^n} \sup_{\mathbb{P}\in\Lambda} \mathbb{P}_n(x) = \sum_{m=0}^{n}\binom{n}{m} \sum_{z\in\{u+1,\ldots\}^m} \sum_{y\in\{1,2,\ldots,u\}^{n-m}} \sup_{\mathbb{P}\in\Lambda} \mathbb{P}_n(zy)$$

$$\leqslant \sum_{m=0}^{n}\binom{n}{m} \sum_{z\in\{u+1,\ldots\}^m} \prod_{i=1}^{m} \sup_{\mathbb{P}\in\Lambda} \mathbb{P}(z_i) \sum_{y\in\{1,2,\ldots,u\}^{n-m}} \sup_{\mathbb{P}\in\Lambda} \mathbb{P}^{n-m}(y)$$

$$\leqslant \left(\sum_{m=0}^{n}\binom{n}{m}\overline{F}_{\Lambda^1}(u)^m\right)\left(\sum_{y\in\{1,2,\ldots,u\}^n} \sup_{\mathbb{P}\in\Lambda} \mathbb{P}_n(y)\right).$$

On the one hand, we have

$$\left(\sum_{m=0}^{n} \binom{n}{m} \overline{F}_{\Lambda^1}(u)^m \right) = \left(1 + \overline{F}_{\Lambda^1}(u)\right)^n.$$

Furthermore,,

$$\log_2 \left(\sum_{\mathbf{y} \in \{1,2,\ldots,u\}^n} \sup_{\mathbb{P} \in \Lambda} \mathbb{P}_n(\mathbf{y}) \right)$$

is the regret of the class of memoryless sources over an alphabet of size u. Since, thanks to Theorem 2.16, for all $\mathbf{y} \in \{1, 2, \ldots, u\}^n$, we have

$$\sup_{\mathbb{P} \in \Lambda} \mathbb{P}_n(\mathbf{y}) \leqslant \mathbb{KT}(\mathbf{y}) \, 2^{\frac{u-1}{2} \log_2 n + 2}$$

we conclude that

$$\sum_{\mathbf{y} \in \{1,2,\ldots,u\}^n} \sup_{\mathbb{P} \in \Lambda} \mathbb{P}_n(\mathbf{y}) \leqslant 2^{\frac{u-1}{2} \log_2 n + 2} \sum_{\mathbf{y} \in \{1,2,\ldots,u\}^n} \mathbb{KT}(\mathbf{y}) = 2^{\frac{u-1}{2} \log_2 n + 2},$$

and obtain

$$R_n^*(\Lambda) \leqslant n \log_2 \left(1 + \overline{F}_{\Lambda}(u)\right) + \frac{u-1}{2} \log_2 n + 2$$

$$\leqslant n \overline{F}_{\Lambda}(u) \log_2 \mathrm{e} + \frac{u-1}{2} \log_2 n + 2.$$

Remark 3.2 The proof of this theorem suggests a coding method by truncation: we encode the truncated part, where we replaced the x_i's strictly larger than u by a specific symbol, with an efficient coding technique in a finite alphabet, for instance mixture coding, and we encode by Elias coding letters x_i's that are strictly larger than u.

Minimax redundancy and regret over an envelop class Λ_f depend on the function f and its behavior at infinity, describing letters' rarity at infinity. We investigate two types of decay: polynomial decay and exponential decay.

3.4.2 Polynomial Decay

Denote, for $\alpha > 1$,

$$\zeta(\alpha) = \sum_{k \geqslant 1} \frac{1}{k^{\alpha}}.$$

Theorem 3.6 *Let $\alpha > 1$ and C such that $C\zeta(\alpha) > 2^\alpha$. Let $\Lambda_{C,-\alpha}$ be the envelop class with polynomial decay given by $f_{\alpha,C} : x \mapsto 1 \wedge \frac{C}{x^\alpha}$. Then,*

$$n^{1/\alpha} A(\alpha) \log_2 \lfloor (C\zeta(\alpha))^{1/\alpha} \rfloor \leqslant \overline{R}_n(\Lambda_{C,-\alpha})$$

with

$$A(\alpha) = \frac{1}{\alpha} \int_1^\infty \frac{1}{u^{1-1/\alpha}} (1 - e^{-1/(\zeta(\alpha)u)}) du,$$

and

$$R_n^*(\Lambda_{C,-\alpha}) \leqslant \left(\frac{2Cn}{\alpha - 1} \right)^{1/\alpha} (\log_2 n)^{1-1/\alpha} + O(1).$$

Remark 3.3 Since the first publication of this book, the upper bound in Theorem 3.6 has been improved by Acharya et al. [3] to

$$R_n^*(\Lambda_{C,-\alpha}) \leqslant (cn)^{1/\alpha} \left(\frac{\alpha}{2} + \frac{1}{\alpha - 1} + \log_2 3 \right) + 1,$$

entailing that the asymptotic order of growth of both minimax redundancy and regret of polynomial envelop classes is $n^{1/\alpha}$.

Let us note that if $\alpha \to +\infty$ and $C = H^\alpha$, $\Lambda_{C,-\alpha}$ converges to the set of memoryless sources over $\{1, \ldots, H\}$ for which the minimax regret is of order $\frac{1}{2}(H-1)\log_2 n$, which is obtained (up to a factor 2) by taking the limit in the upper bound of Theorem 3.6. On the other hand, the limit of the lower bound when α tends to 1 is infinite.

Proof (Proof of Theorem 3.6). For the upper bound,

$$\overline{F}_{\alpha,C}(u) = \sum_{k > u} 1 \wedge \frac{C}{k^\alpha} \leqslant \frac{C}{(\alpha - 1) u^{\alpha - 1}}.$$

Thus, choosing $u_n = \left(\frac{2Cn}{(\alpha-1)\log_2 n} \right)^{\frac{1}{\alpha}}$, we obtain, by Theorem 3.5:

$$R_n^*(\Lambda_{C,-\alpha}) \leqslant \left(\frac{2Cn}{\alpha - 1} \right)^{1/\alpha} (\log_2 n)^{1-1/\alpha} + O(1).$$

For the lower bound, let p and m be two integers such that $m^\alpha \leqslant C\zeta_p(\alpha)$, where

$$\zeta_p(\alpha) = \sum_{k=1}^p \frac{1}{k^\alpha}.$$

Let $W = (\theta_1, \theta_2, \ldots, \theta_p)$, where $(\theta_k)_{1 \leqslant k \leqslant p}$ is a sequence of i.i.d. uniform random variables over $\{1, \ldots, m\}$. For all $W = (\theta_1, \theta_2, \ldots, \theta_p)$, let P_W be the probability over $\mathbb{N}$ with support $\bigcup_{1 \leqslant k \leqslant p} \{(k-1)m + \theta_k\}$, such that

$$P_W\big((k-1)m+\theta_k\big) = \frac{1}{\zeta_p(\alpha)k^\alpha} = \frac{m^\alpha}{\zeta_p(\alpha)} \cdot \frac{1}{(k\,m)^\alpha} \quad \text{for } 1 \leqslant k \leqslant p. \qquad (3.2)$$

Condition $m^\alpha \leqslant C\zeta_p(\alpha)$ gives $P_W^{\otimes\mathbb{N}} \in \Lambda_{\alpha,C}$. Let $X_1, \ldots, X_n$ be random variables which, conditionally on W, are independent and identically distributed according to P_W. By Theorem 2.12

$$\overline{R}_n(\Lambda_{C\bullet^{-\alpha}}) \geqslant I\,(W, X_{1:n})\,.$$

But

$$I\,(W, X_{1:n}) = H\,(W) - H\,(W\,|\,X_{1:n}) = \sum_{k=1}^{p} H\,(\theta_k) - H\,(W\,|\,X_{1:n})$$

$$\geqslant \sum_{k=1}^{p} \big(H\,(\theta_k) - H\,(\theta_k\,|\,X_{1:n})\big)\,.$$

For $1 \leqslant k \leqslant p$, let $N_k\,(X_{1:n}) = 1$ if there exists an index $i \in \{1, \ldots, n\}$ such that $X_i \in [(k-1)m+1, km]$ and 0 otherwise. If $N_k\,(x_{1:n}) = 0$, the conditional distribution of θ_k given $X_{1:n} = x_{1:n}$ is uniform over $\{1, \ldots, m\}$. If $N_k\,(x_{1:n}) = 1$, the conditional distribution of θ_k given $X_{1:n} = x_{1:n}$ is Dirac (the value θ_k is known). Hence

$$H\,(\theta_k\,|\,X_{1:n}) = \sum_{x_{1:n}} \mathbb{P}\,(x_{1:n})\,H\,(\theta_k\,|\,X_{1:n} = x_{1:n}) = \mathbb{P}(N_k = 0)\log_2 m$$

so that

$$I\,(W, X_{1:n}) \geqslant \sum_{k \geqslant 1} \mathbb{P}(N_k = 1)\log_2 m = E\,[Z_n]\log_2 m,$$

where $Z_n\,(X_{1:n})$ is the number of distinct letters in $X_{1:n}$ (its law does not depend on W). We have

$$E\,[Z_n] = \sum_{k=1}^{p} \left(1 - \left(1 - \frac{1}{\zeta_p(\alpha)k^\alpha}\right)^n\right)$$

thus

$$\overline{R}_n(\Lambda_{C\bullet^{-\alpha}}) \geqslant \left(\sum_{k=1}^{p} \left(1 - \left(1 - \frac{1}{\zeta_p(\alpha)k^\alpha}\right)^n\right)\right)\log_2 m$$

and letting $p \to +\infty$, we find that for all m such that $m^\alpha \leqslant C\zeta(\alpha)$

$$\overline{R}_n(\Lambda_{C\bullet^{-\alpha}}) \geqslant \left(\sum_{k=1}^{+\infty} \left(1 - \left(1 - \frac{1}{\zeta(\alpha)k^\alpha}\right)^n\right)\right)\log_2 m.$$

Now,

$$\sum_{k=1}^{\infty} \left(1 - \left(1 - \frac{1}{\zeta(\alpha)k^{\alpha}}\right)^n\right) \geqslant \sum_{k=1}^{\infty} \left(1 - \exp\left(-\frac{n}{\zeta(\alpha)\,k^{\alpha}}\right)\right)$$

$$\text{since } 1 - x \leqslant \exp(-x)$$

$$\geqslant \int_1^{\infty} \left(1 - \exp\left(-\frac{n}{\zeta(\alpha)x^{\alpha}}\right)\right) dx$$

$$\geqslant \frac{n^{\frac{1}{\alpha}}}{\alpha} \int_1^{\infty} \frac{1}{u^{1-\frac{1}{\alpha}}} \left(1 - \exp\left(-\frac{1}{\zeta(\alpha)u}\right)\right) du.$$

We then choose m as large as possible, namely $m = \lfloor (C\zeta(\alpha))^{1/\alpha} \rfloor$.

3.4.3 Exponential Decay

Theorem 3.7 *Let C and α be non-negative real numbers such that $C > e^{2\alpha}$. Let $\Lambda_{Ce^{-\alpha}}$ be the envelop class with exponential decay given by the function $f_{\alpha} : x \mapsto 1 \wedge C\,e^{-\alpha x}$. Then*

$$\frac{1}{8\alpha}\left(\log_2 n\right)^2 (1 - o(1)) \leqslant \overline{R}_n(\Lambda_{Ce^{-\alpha\cdot}}) \leqslant R_n^*(\Lambda_{Ce^{-\alpha\cdot}}) \leqslant \frac{1}{2\alpha}(\log_2 n)^2 + O(1).$$

Remark 3.4 In this case, the speed order is known and is given by $\frac{1}{\alpha} \log_2^2 n$. D. Bontemps [4] showed that the redundancy is asymptotically equivalent to $\frac{1}{4\alpha e}(\log_2 n)^2$. Its proof uses sophisticated results of Haussler and Oper [5].

Proof (Proof of Theorem 3.7). For the upper bound,

$$\overline{F}_{\alpha}(u) = \sum_{k>u} 1 \wedge C\,e^{-\alpha k} \leqslant \frac{C}{1 - e^{-\alpha}}\,e^{-\alpha(u+1)}.$$

We choose $u_n = \frac{1}{\alpha}\log_2 n$ and obtain, by Theorem 3.5,

$$R_n^*(\Lambda_{Ce^{-\alpha\cdot}}) \leqslant \frac{1}{2\alpha}(\log_2 n)^2 + O(1).$$

For the lower bound, let $p = \lfloor \frac{1}{2\alpha}\log_2 \frac{nC}{4\log_2 n} \rfloor$, so that we have $n\,e^{-2\alpha p} \geqslant 4\log_2 n/C$ for n large enough. For $1 \leqslant j \leqslant p$, let

$$u_j = \left\lfloor \sqrt{\frac{n(C\,e^{-2\alpha j} \wedge 1))}{\log_2 n}} \right\rfloor.$$

Note that $u_p = 2$.

Let
$$\Theta = \left\{ (\theta_1, \ldots, \theta_p) : \forall j, \; \theta_j \in \left\{ \frac{1}{u_j}, \frac{2}{u_j}, \cdots, 1 \right\} \right\}.$$

Let
$$c_p = \sum_{j=1}^{p} e^{-2\alpha j} = e^{-2\alpha} \frac{1 - e^{-2\alpha p}}{1 - e^{-2\alpha}},$$

so that we have $c_p > 1/C$ for p large enough. Let W have uniform distribution over Θ and $X_1, \ldots, X_n$ be random variables which, conditionally on $W = \theta$, are independent and identically distributed according to p_θ over $\{1, \ldots, 2p\}$ such that for $1 \leqslant j \leqslant p$:

- $p_\theta(2j - 1) = \theta_j e^{-2\alpha j}/c_p$;
- $p_\theta(2j) = (1 - \theta_j) e^{-2\alpha j}/c_p$.

We have $p_\theta^{\otimes \mathbb{N}} \in \Lambda_{C e^{-\alpha \bullet}}$. Hence

$$\bar{R}_n(\Lambda_{C e^{-\alpha \bullet}}) \geqslant I(W; X_{1:n}) = H(W) - H(W \mid X_{1:n}).$$

Now,

$$
\begin{aligned}
H(W) &= \log_2 |\Theta| \\
&= \sum_{j=1}^{p} \log_2 \left\lfloor \sqrt{n(C e^{-2\alpha j} \wedge 1)/\log_2 n} \right\rfloor \\
&= \sum_{j=1}^{p} \log_2 \sqrt{n(C e^{-2\alpha j} \wedge 1)/\log_2 n} + O(\log_2 n) \\
&= \frac{p}{2} \left(\log_2 n - \log_2 \log_2 n \right) - \frac{1}{2} \sum_{j=a}^{p} \log_2 (C e^{-2\alpha j}) + O(\log_2 n) \\
&= \frac{p}{2} \left(\log_2 n - \log_2 \log_2 n \right) - \frac{(p - a + 1) \log_2 C}{2} \\
&\quad - \frac{\alpha}{2} \left((p(p+1)) - a(a - 1) \right) + O(\log_2 n) \\
&= \frac{1}{8\alpha} (\log_2 n)^2 + O(\log_2 n).
\end{aligned}
$$

We take advantage of the information about W contained in $X_{1:n}$ to evaluate $H(W \mid X_{1:n})$. If $\widehat{\theta}$ is a function of $X_{1:n}$ (an "estimator" of W), we have, by Fano's Inequality (Theorem 3.8, proved at the end of the section)

$$H(W \mid X_{1:n}) \leqslant P(W \neq \widehat{\theta}) \log_2 |\Theta| + 1$$

and it suffices to show that $P(W \neq \widehat{\theta}) = o(1)$ to obtain the desired result.

Let, for $1 \leqslant i \leqslant 2p$ and $1 \leqslant j \leqslant p$,

$$N_i = \sum_{k=1}^{n} \mathbf{1}_{X_k = i} \quad \text{and} \quad Z_j = N_{2j-1} + N_{2j}.$$

Let then A_j be the integer closest to $u_j N_{2j-1}/Z_j$. Define $\widehat{\theta} = (\widehat{\theta}_j)_{1 \leqslant j \leqslant p}$ by:

$$\widehat{\theta}_j = \frac{A_j}{u_j}.$$

We then have

$$P(\widehat{\theta} \neq W) = \frac{1}{|\Theta|} \sum_{\theta \in \Theta} \mathbb{P}(\widehat{\theta} \neq \theta \mid W = \theta)$$

and

$$\mathbb{P}(\widehat{\theta} \neq \theta \mid W = \theta) \leqslant \sum_{j=1}^{p} \mathbb{P}(\widehat{\theta}_j \neq \theta_j \mid W = \theta).$$

Conditionally on W, Z_j has distribution $\mathscr{B}(n, \mathrm{e}^{-2\alpha j}/c_p)$, with variance upper-bounded by $n\,\mathrm{e}^{-2\alpha j}/c_p$. By Bernstein's Inequality (recalled at the end of the section):

$$\mathbb{P}\left(Z_j < \frac{n\,\mathrm{e}^{-2\alpha j}}{2c_p} \mid W = \theta\right) \leqslant \exp\left(-\frac{1}{2} \cdot \frac{(n\,\mathrm{e}^{-2\alpha j}/2c_p)^2}{n\,\mathrm{e}^{-2\alpha j}/c_p + n\,\mathrm{e}^{-2\alpha j}/6c_p}\right)$$

$$\leqslant \exp\left(-\frac{n\,\mathrm{e}^{-2\alpha j}}{10c_p}\right)$$

$$\leqslant \exp\left(-\frac{n\,\mathrm{e}^{-2\alpha p}}{10c_p}\right) \leqslant \exp\left(-\frac{4\log_2 n}{10Cc_p}\right) \leqslant n^{-1/3Cc_p}.$$

Also, conditionally on W and Z_j, N_{2j-1} has distribution $\mathscr{B}(Z_j, \theta_j)$. By Hoeffding's Inequality (recalled at the end of the section):

$$\mathbb{P}(\widehat{\theta}_j \neq \theta_j \mid W = \theta, Z_j = z) \leqslant \mathbb{P}\left(\left|\frac{N_{2j-1}}{z} - \theta_j\right| > \frac{1}{2u_j} \mid Z_j = z, W = \theta\right)$$

$$\leqslant 2\exp\left(-\frac{2z}{4u_j^2}\right)$$

$$\leqslant 2\exp\left(-\frac{z\log_2 n}{2n(C\,\mathrm{e}^{-2\alpha j} \wedge 1)}\right) \leqslant 2\exp\left(-\frac{z\log_2 n}{2nC\,\mathrm{e}^{-2\alpha j}}\right).$$

So if $z \geqslant n\,\mathrm{e}^{-2\alpha j}/2c_p$:

$$\mathbb{P}\left(\left|\frac{N_{2j-1}}{z} - \theta_j\right| > \frac{1}{2u_j} \,\Big|\, Z_j = z, W = \theta\right) \leqslant 2\exp\left(-\frac{\frac{ne^{-2\alpha j}}{2c_p}\log_2 n}{2nC\,e^{-2\alpha j}}\right) \leqslant n^{-\frac{1}{4Cc_p}}$$

and thus

$$\mathbb{P}(\widehat{\theta}_j \neq \theta_j \,|\, W = \theta)$$

$$\leqslant \mathbb{P}\left(Z_j < \frac{ne^{-2\alpha j}}{2c_p} \,\Big|\, W = \theta\right)$$

$$+ \sum_{z=\lceil ne^{-2\alpha j}/2c_p\rceil}^{n} \mathbb{P}\left(\left|\frac{N_{2j-1}}{z} - \theta_j\right| > \frac{1}{2u_j} \,\Big|\, \begin{matrix} Z_j = z \\ W = \theta \end{matrix}\right) \times \mathbb{P}(Z_j = z \,|\, W = \theta)$$

$$\leqslant n^{-\frac{1}{3Cc_p}} + n^{-\frac{1}{4Cc_p}}$$

so

$$\mathbb{P}(\widehat{\theta} \neq \theta \,|\, W = \theta) \leqslant p\left(n^{-\frac{1}{3Cc_p}} + n^{-\frac{1}{4Cc_p}}\right) = o(1).$$

The result follows.

Let us now recall the three fundamental inequalities that were used in the proof.

Theorem 3.8 (Fano's Inequality) *Let Θ be a finite set, W and $\widehat{\theta}$ be two random variables with values in Θ and $p_e = P(\widehat{\theta} \neq W)$. Then*

$$H(W \,|\, \widehat{\theta}) \leqslant h(p_e) + p_e \log_2\left(|\Theta| - 1\right),$$

where $h(x) = -x\log_2 x - (1-x)\log_2(1-x)$.

Proof (Proof of Theorem 3.8). Let $T = \mathbf{1}_{\widehat{\theta} \neq W}$. We have

$$H\big((W, T)\,|\,X\big) = H(W \,|\, X) + H\big(T \,|\, (W, X)\big) = H(T \,|\, X) + H\big(W \,|\, (T, X)\big).$$

But $H(T \,|\, (W, X)) = 0$ and $H(T \,|\, X) \leqslant H(T) = h(p_e)$. Also,

$$H\big(W \,|\, (T, X)\big) = (1 - p_e)H(W \,|\, T = 0, X) + p_e H(W \,|\, T = 1, X).$$

Given $T = 0$, $W = g(X)$ so $H\,(W \,|\, T = 0, X) = 0$. Given $T = 1$, $W \neq g(X)$ so W takes values in a set of size at most $|\Theta| - 1$, hence $H\,(W \,|\, T = 1, X) \leqslant \log_2(|\Theta| - 1)$.

Theorem 3.9 (Bernstein's Inequality) *Let $Y_1, \ldots, Y_n$ be independent, real-valued, bounded, centered random variables such that Y_i has values in $[-M, M]$, for $i = 1, \ldots, n$. Then, for all real numbers $u > 0$,*

$$\mathbb{P}\left(\sum_{i=1}^{n} Y_i \geqslant u\right) \leqslant \exp\left(-\frac{1}{2} \cdot \frac{u^2}{v + \frac{1}{3}Mu}\right),$$

where $v \geqslant Var(\sum_{i=1}^{n} Y_i)$.

One may find a proof of Theorem 3.9 in [6].

Theorem 3.10 (Hoeffding's Inequality) *Let* $Y_1, \ldots, Y_n$ *be independent, real-valued, bounded, centered random variables such that, for* $i = 1, \ldots, n$, Y_i *has values in* $[a_i, b_i]$. *Then, for all real numbers* $u > 0$,

$$\mathbb{P}\left(\sum_{i=1}^{n} Y_i \geqslant u\right) \leqslant \exp\left(-\frac{2u^2}{\sum_{i=1}^{n}(b_i - a_i)^2}\right).$$

One may find a proof of Theorem 3.10 in [6].

3.5 Patterns

Another possible approach to code sources with values in large alphabets is to separate coding of the observed dictionary on the one hand, and coding of the "pattern" on the other hand (i.e. order of occurrence and repetitions of letters of the observed dictionary). For instance, if the word is $x_{1:11} = abracadabra$, the observed dictionary is $\{a; b; r; c; d\}$ and the pattern is 12314151231. Note that to retrieve the coded word, it is important that the dictionary be ordered according to the order of occurrence of letters in the word. In other words, although the dictionary is defined as a set, we endow it with an order according to the occurrence of letters in the word, and encoding of the dictionary is done in that order.

Formally, for all $x_{1:n} \in \mathscr{X}^n$, the dictionary of word $x_{1:n}$ is the set

$$\mathscr{A}(x_{1:n}) = \{x_1, \ldots, x_n\}$$

and the pattern $\Psi(x_{1:n}) = \Psi_{1:n}$ gives the rank of arrival Ψ_i of letter x_i in the word. More precisely, if we define for all x in $\mathscr{A}(x_{1:n})$,

$$\ell_{x_{1:n}}(x) = \min\left\{|\mathscr{A}(x_{1:i})| : x_i = x\right\},$$

then $\Psi_{1:n} = \ell_{x_{1:n}}(x_1) \ldots \ell_{x_{1:n}}(x_n)$. Note that the function $\ell_{x_{1:n}}(\cdot)$ is a one-to-one map from $\mathscr{A}(x_{1:n})$ into $\{1, \ldots, |\mathscr{A}(x_{1:n})|\}$, and that this one-to-one map orders $\mathscr{A}(x_{1:n})$ according to the order of occurrence of letters in word $x_{1:n}$.

Denoting by Ψ^n the set of possible patterns of length n, we have

$$\Psi^1 = \{1\}, \quad \Psi^2 = \{11, 12\}, \quad \Psi^3 = \{111, 112, 121, 122, 123\}, \ldots$$

and Ψ^n is a strict subset of $\{1, \ldots, n\}^n$.

If now $\mathbb{P}$ is the distribution of a stationary source $(X_n)_{n \in \mathbb{N}}$, the induced distribution on patterns is the image-distribution $\mathbb{P}_\Psi$, which is not stationary anymore. Indeed, for all integers n, the support of $\mathbb{P}_{\Psi}^n$, the distribution of $\Psi_{1:n}^{(n)}$, is not of the form S^n for some $S \subset \mathcal{X}$. Let n be an integer, and let ϕ be a uniquely decodable code over Ψ^n. We always have

$$E\big[\ell\big(\phi(\Psi(X_{1:n}))\big)\big] \geqslant H(\mathbb{P}_{\Psi}^n),$$

but what happens when n tends to infinity? In particular, does there still exist an entropy rate for the distribution $\mathbb{P}_{\Psi}$?

3.5.1 Entropy Rate of Patterns

The following result is due to Gemelos and Weissmann [7].

Theorem 3.11 *If* $(X_n)_{n \in \mathbb{N}}$ *is a stationary ergodic source with distribution* $\mathbb{P}$, *values in* $\mathcal{X}$ *countable, and finite entropy rate* $H_*(\mathbb{P})$, *then*

$$\lim_{n \to +\infty} \frac{1}{n} H(\mathbb{P}_{\Psi}^n) = H_*(\mathbb{P}).$$

Proof (Proof of Theorem 3.11). We extend the stationary ergodic process $(X_n)_{n \in \mathbb{N}}$ to a stationary ergodic process $(X_n)_{n \in \mathbb{Z}}$ and start by showing that if $\Psi(X_{-n:0}) = \Psi_{-n:0}^{(n)}$, then

$$\lim_{n \to +\infty} E\big[-\log_2 \mathbb{P}(\Psi_0^{(n)} \mid X_{-n:-1}) \big] = H^\infty = H_*(\mathbb{P}),$$

where $H^\infty = E\big[-\log_2 \mathbb{P}(X_0 \mid X_{-\infty:-1}) \big]$. Since

$$H(X_0, \Psi_0^{(n)} \mid X_{-n:-1}) = H(X_0 \mid X_{-n:-1}) + H(\Psi_0^{(n)} \mid X_{-n:0})$$
$$= H(\Psi_0^{(n)} \mid X_{-n:-1}) + H(X_0 \mid \Psi_0^{(n)}, X_{-n:-1})$$

and $H(\Psi_0^{(n)} \mid X_{-n:0}) = 0$ because $\Psi_0^{(n)}$ is a function of $X_{-n:0}$, we have

$$H(\Psi_0^{(n)} \mid X_{-n:-1}) \leqslant H(X_0 \mid X_{-n:-1}).$$

Let us now show that

$$\limsup_{n \to +\infty} \big[H(X_0 \mid X_{-n:-1}) - H(\Psi_0^{(n)} \mid X_{-n:-1}) \big] \leqslant 0.$$

We have

$$H(\Psi_0^{(n)} \mid X_{-n:-1}) = E\bigg\{ \sum_{x \in \mathscr{A}(X_{-n:-1})} -\mathbb{P}(X_0 = x \mid X_{-n:-1}) \log_2 \mathbb{P}(X_0 = x \mid X_{-n:-1})$$

$$- \mathbb{P}(X_0 \notin \mathscr{A}(X_{-n:-1}) \mid X_{-n:-1}) \log_2 \mathbb{P}(X_0 \notin \mathscr{A}(X_{-n:-1}) \mid X_{-n:-1}) \bigg\}$$

and thus

$$H\left(X_0 \mid X_{-n:-1}\right) - H(\Psi_0^{(n)} \mid X_{-n:-1})$$
$$= E\bigg\{ \sum_{x \notin \mathscr{A}(X_{-n:-1})} -\mathbb{P}\left(X_0 = x \mid X_{-n:-1}\right) \log_2 \mathbb{P}\left(X_0 = x \mid X_{-n:-1}\right)$$
$$+ \mathbb{P}(X_0 \notin \mathscr{A}(X_{-n:-1}) \mid X_{-n:-1}) \log_2 \mathbb{P}(X_0 \notin \mathscr{A}(X_{-n:-1}) \mid X_{-n:-1}) \bigg\}$$
$$\leqslant E\bigg\{ \sum_{x \notin \mathscr{A}(X_{-n:-1})} -\mathbb{P}(X_0 = x \mid X_{-n:-1}) \log_2 \mathbb{P}(X_0 = x \mid X_{-n:-1}) \bigg\}.$$

Let $\varepsilon > 0$. $H(X_1)$ is finite, so there exists a finite set B such that for all $x \in B$, $\mathbb{P}(X_0 = x) > 0$, and

$$-\mathbb{P}(X_0 \in B) \log_2 \mathbb{P}(X_0 \in B) - \sum_{x \notin B} \mathbb{P}(X_0 = x) \log_2 \mathbb{P}(X_0 = x) \leqslant \varepsilon.$$

In other words, if Z is the random variable equal to X_0 if $X_0 \notin B$ and an arbitrary fixed value (but not an element of $\mathscr{X}$) if $X_0 \in B$, then $H(Z) \leqslant \varepsilon$.

Now, we split, in the expectation of the previous upper bound, according to whether $B \subset \mathscr{A}(X_{-n:-1})$ or not, so that

$$H(X_0 \mid X_{-n:-1}) - H(\Psi_0^{(n)} \mid X_{-n:-1})$$
$$\leqslant E\bigg\{ \sum_{x \notin B} -\mathbb{P}(X_0 = x \mid X_{-n:-1}) \log_2 \mathbb{P}(X_0 = x \mid X_{-n:-1}) \bigg\}$$
$$+ E\bigg\{ \Big(\sum_{x \in \mathscr{X}} -\mathbb{P}\left(X_0 = x \mid X_{-n:-1}\right) \log_2 \mathbb{P}(X_0 = x \mid X_{-n:-1}) \Big)(1 - \mathbf{1}_{B \subset \mathscr{A}(X_{-n:-1})}) \bigg\}.$$

But

$$E\bigg\{ \sum_{x \notin B} -\mathbb{P}(X_0 = x \mid X_{-n:-1}) \log_2 \mathbb{P}(X_0 = x \mid X_{-n:-1}) \bigg\}$$
$$= E\left\{ H(Z \mid X_{-n:-1}) + \mathbb{P}(X_0 \in B \mid X_{-n:-1}) \log_2 \mathbb{P}(X_0 \in B \mid X_{-n:-1}) \right\}$$
$$\leqslant H(Z) \leqslant \varepsilon$$

and

$$E\bigg\{ \Big(\sum_{x \in \mathscr{X}} -\mathbb{P}(X_0 = x \mid X_{-n:-1}) \log_2 \mathbb{P}(X_0 = x \mid X_{-n:-1}) \Big)(1 - \mathbf{1}_{B \subset \mathscr{A}(X_{-n:-1})}) \bigg\}$$
$$= E\left\{ H(X_0 \mid X_{-n:-1}) \left(1 - \mathbf{1}_{B \subset \mathscr{A}(X_{-n:-1})}\right) \right\}$$
$$\leqslant E\left\{ H(X_0) \left(1 - \mathbf{1}_{B \subset \mathscr{A}(X_{-n:-1})}\right) \right\} = H(X_0) \left[1 - \mathbb{P}(B \subset \mathscr{A}(X_{-n:-1}))\right]$$

so that

$$H\left(X_0 \mid X_{-n:-1}\right) - H(\Psi_0^{(n)} \mid X_{-n:-1}) \leqslant \varepsilon + H(X_0) \sum_{x \in B} \mathbb{P}(x \notin \mathscr{A}(X_{-n:-1})).$$

Now by Theorem 1.9, for all $x \in B$, the event

$$\left(x \in \mathscr{A}(X_{-n:-1})\right) = \left(\sum_{k=1}^{n} \mathbf{1}_{X_{-k}=x} \geqslant 1\right)$$

has probability tending to 1 as n tends to infinity, since $\mathbb{P}(X_1 = x) > 0$. Hence, B being finite, $\sum_{x \in B} \mathbb{P}(x \notin \mathscr{A}(X_{-n:-1}))$ tends to 0 as n tends to infinity, and finally, for all $\varepsilon > 0$,

$$\limsup_{n \to +\infty} \left\{ H\left(X_0 \mid X_{-n:-1}\right) - H(\Psi_0^{(n)} \mid X_{-n:-1}) \right\} \leqslant \varepsilon.$$

We may then conclude that

$$\lim_{n \to +\infty} H(\Psi_0^{(n)} \mid X_{-n:-1}) = H_*\left(\mathbb{P}\right).$$

Now,

$$\frac{1}{n} H\left(X_{1:n}\right) \geqslant \frac{1}{n} H\left(\Psi(X_{1:n})\right)$$

since

$$H\left(X_{1:n}, \Psi(X_{1:n})\right) = H\left(X_{1:n}\right) + H\left(\Psi(X_{1:n}) \mid X_{1:n}\right) = H\left(\Psi(X_{1:n})\right) + H\left(X_{1:n} \mid \Psi(X_{1:n})\right)$$

and $H\left(\Psi(X_{1:n}) \mid X_{1:n}\right) = 0$, so that

$$\limsup_{n \to +\infty} \frac{1}{n} H\left(\Psi(X_{1:n})\right) \leqslant H_*(\mathbb{P}).$$

In order to conclude the proof, we show that $\liminf_{n \to +\infty} \frac{1}{n} H\left(\Psi(X_{1:n})\right)) \geqslant H_*\left(\mathbb{P}\right)$. To do so, we use the stationarity of $(X_n)_{n \in \mathbb{Z}}$, allowing us to write

$$\frac{1}{n} H\left(\Psi(X_{1:n})\right) = \frac{1}{n} \sum_{i=1}^{n} H\left(\Psi(X_{1:n})_i \mid \Psi(X_{1:n})_{1:i-1}\right) = \frac{1}{n} \sum_{i=1}^{n} H(\Psi_0^{(n)} \mid \Psi_{-i:-1}^{(n)}).$$

But for all $i = 1, \ldots, n$,

$$H(\Psi_0^{(n)} \mid \Psi_{-i:-1}^{(n)}) \geqslant H(\Psi_0^{(n)} \mid \Psi_{-n:-1}^{(n)})$$

so that

$$\frac{1}{n}H\big(\Psi(X_{1:n})\big) \geqslant H(\Psi_0^{(n)} \,|\, \Psi_{-n:-1}^{(n)}).$$

Now,

$$
\begin{aligned}
H(\Psi_0^{(n)}, X_{-n:-1} \,|\, \Psi_{-n:-1}^{(n)}) &= H(\Psi_0^{(n)} \,|\, \Psi_{-n:-1}^{(n)}) + H(X_{-n:-1} \,|\, \Psi_{-n:0}^{(n)}) \\
&= H(X_{-n:-1} \,|\, \Psi_{-n:-1}^{(n)}) + H(\Psi_0^{(n)} \,|\, \Psi_{-n:-1}^{(n)}, X_{-n:-1}) \\
&= H(X_{-n:-1} \,|\, \Psi_{-n:-1}^{(n)}) + H(\Psi_0^{(n)} \,|\, X_{-n:-1})
\end{aligned}
$$

thus

$$H(\Psi_0^{(n)} \,|\, \Psi_{-n:-1}^{(n)}) - H(\Psi_0^{(n)} \,|\, X_{-n:-1}) = H(X_{-n:-1} \,|\, \Psi_{-n:-1}^{(n)}) - H(X_{-n:-1} \,|\, \Psi_{-n:0}^{(n)}) \geqslant 0,$$

which implies

$$\frac{1}{n}H\big(\Psi(X_{1:n})\big) \geqslant H(\Psi_0^{(n)} \,|\, X_{-n:-1})$$

so that

$$\liminf_{n\to+\infty} \frac{1}{n}H\big(\Psi(X_{1:n})\big) \geqslant \lim_{n\to+\infty} H(\Psi_0^{(n)} \,|\, X_{-n:-1}) = H_*\,(\mathbb{P}).$$

3.5.2 Regrets and Redundancies of Patterns

For a set of stationary ergodic sources Λ, let us define regrets and redundancy of patterns:

$$R_n^{*,\Psi}\,(\Lambda) = \inf_{Q_n} \sup_{\mathbb{P}\in\Lambda} \sup_{\Psi_{1:n}\in\Psi^n} \big\{ \log_2 \mathbb{P}_\Psi^n\,(\Psi_{1:n}) - \log_2 Q_n\,(\Psi_{1:n}) \big\},$$

$$\overline{R}_n^{\Psi}\,(\Lambda) = \inf_{Q_n} \sup_{\mathbb{P}\in\Lambda} D(\mathbb{P}_\Psi^n \,|\, Q_n),$$

where in both cases the infimum is taken over probability distributions Q_n over Ψ^n. The following upper bound is due to Orlitsky and Santhanam [8].

Theorem 3.12 *If Λ is the set of memoryless sources, then*

$$R_n^{*,\Psi}(\Lambda) \leqslant \left(\pi\sqrt{\frac{2}{3}}\,\log_2 e\right)\sqrt{n}.$$

Proof (Proof of Theorem 3.12). Given a sequence of letters $z_{1:n}$, its profile is given by

$$\gamma(z_{1:n}) = (\gamma_1, \ldots, \gamma_n)$$

where, for each $i = 1, \ldots, n$, γ_i is the number of letters occurring i times in $z_{1:n}$. We always have $\sum_{i=1}^{n} i\gamma_i = n$. We let Γ^n be the set of possible profiles of patterns:

$$\Gamma^n = \{\gamma(\Psi_{1:n}), \ \Psi_{1:n} \in \Psi^n\}.$$

Also, for $\gamma \in \Gamma^n$, we let:

$$\Psi_\gamma = \{\Psi_{1:n} \in \Psi^n, \ \gamma(\Psi_{1:n}) = \gamma\}.$$

Note that $\mathbb{P}_\Psi(\psi_{1:n})$ is constant over Ψ_γ, so for all $\varepsilon > 0$, if $\widehat{\mathbb{P}}$ satisfies

$$\sup_{\mathbb{P} \in \Lambda} \mathbb{P}_\Psi(\psi_{1:n}) \leqslant (1+\varepsilon)\widehat{\mathbb{P}}(\Psi_{1:n}),$$

then we can always take the same $\widehat{\mathbb{P}}$ for all $\Psi_{1:n}$ in Ψ_γ and we thus have

$$R_n^{*,\Psi}(\Lambda) = \log_2 \sum_{\gamma \in \Gamma^n} \sum_{\Psi_{1:n} \in \Psi_\gamma} \sup_{\mathbb{P} \in \Lambda} \mathbb{P}_\Psi(\psi_{1:n}) \leqslant \log_2(1+\varepsilon) + \log_2 |\Gamma^n|.$$

Note that $\gamma \in \Gamma^n$ can be interpreted as a non-ordered partition of integer n, i.e. a collection of integers summing to n, the integer γ_i corresponding to the number of times integer i occurs in the partition. The upper bound then follows from the following result of Hardy and Ramanujan [9]:

$$\left(\left(\pi\sqrt{\frac{2}{3}}\log_2 e\right)\sqrt{n}\right)(1 - o(1)) \leqslant \log_2 |\Gamma^n| \leqslant \left(\pi\sqrt{\frac{2}{3}}\log_2 e\right)\sqrt{n} \qquad \square$$

The following lower bound is due to Garivier [10].

Theorem 3.13 *For all n large enough:*

$$\overline{R}_n^\Psi(\Lambda) \geqslant 1.84\left(\frac{n}{\log_2 n}\right)^{1/3}.$$

Proof (Proof of Theorem 3.13). Let $c = c_n$ be an integer, $\lambda \in \mathbb{R}^+$ and $d = \lambda\sqrt{c}$. Let $\Theta^{c,d}$ be the set

$$\Theta^{c,d} = \left\{\theta = (\theta_j)_{1 \leqslant j \leqslant c} : d \geqslant \theta_1 \geqslant \theta_2 \geqslant \cdots \geqslant \theta_c \text{ and } \sum_{j=1}^{c} \theta_j = c\right\}.$$

For $\theta \in \Theta^{c,d}$, let p_θ be the probability over $\{1, \ldots, c\}$ given by $p_\theta(i) = \theta_i/c$ and $\mathbb{P}_\theta$ be the distribution of a memoryless source with marginal p_θ. Let W be a uniform random variable over $\Theta^{c,d}$ and $(X_n)_{n \geqslant 1}$ the source with distribution $\mathbb{P}_\theta$ conditionally on $W = \theta$. By Theorem 2.12

$$\overline{R}_n^\Psi (\Lambda) \geqslant I\left(W;\, \Psi(X_{1:n})\right).$$

We have

$$I\left(W;\, \Psi(X_{1:n})\right) = H\left(W\right) - H\left(W \mid \Psi(X_{1:n})\right)$$

and

$$H\left(W\right) = \log_2 \Theta^{c,d} = (\log_2 e) f(\lambda) \sqrt{c}\left(1 + o(1)\right)$$

by Dixmier and Nicolas [11] for some tabulated function f. On the other hand, if $\widehat{\theta}$ is a function of $\Psi(X_{1:n})$, then by Fano's Inequality 3.8:

$$H\left(W \mid \Psi(X_{1:n})\right) \leqslant \mathbb{P}\left(W \neq \widehat{\theta}\right) \log_2 \Theta^{c,d} + 1$$

and thus

$$I\left(W;\, \Psi(X_{1:n})\right) \geqslant \log_2 \Theta^{c,d}\left[1 - \mathbb{P}(W \neq \widehat{\theta})\right] - 1.$$

We thus have to find $\widehat{\theta}$ such that $\mathbb{P}(W \neq \widehat{\theta})$ is small. In other words, if $x_{1:n}$ was generated by $p_\theta^{\otimes n}$, we want to recover θ from $\Psi(x_{1:n})$.

Let T_j be the number of occurrences of the jth most frequent symbol in $\Psi_{1:n}$, i.e. if $\widetilde{T}_j = \sum_{i=1}^n \mathbf{1}_{\Psi_i = j}$, $(T_j)_{j=1,\dots,n}$ is the reverse order statistics of $(\widetilde{T}_j)_{j=1,\dots,n}$. With large probability, T_j/n will be close to θ_j/c. We then define $\widehat{\theta} = (\widehat{\theta}_j)_{1 \leqslant j \leqslant c}$, with

$$\widehat{\theta}_j = \left[\frac{T_j c}{n}\right].$$

For $1 \leqslant j \leqslant n$, let $U_j = \sum_{i=1}^n \mathbf{1}_{X_i = j}$. Since the number of occurrences of the jth most frequent symbol in $X_{1:n}$ is still equal to T_j, $(T_j)_{j=1,\dots,n}$ is also the reverse order statistics of $(U_j)_{j=1,\dots,n}$. We have

$$\{\widehat{\theta} = \theta\} \supset \bigcap_{j=1}^c \left\{\left|\frac{T_j c}{n} - \theta_j\right| < \frac{1}{2}\right\},$$

and thus

$$\{\widehat{\theta} = \theta\} \supset \bigcap_{j=1}^c \left\{\left|\frac{U_j c}{n} - \theta_j\right| < \frac{1}{2}\right\} \cap \{U_1 \geqslant U_2 \geqslant \cdots \geqslant U_c\}.$$

If all θ_j's are distinct, then

$$\{U_1 \geqslant U_2 \geqslant \cdots \geqslant U_c\} \subset \bigcap_{j=1}^c \left\{\left|\frac{U_j c}{n} - \theta_j\right| < \frac{1}{2}\right\}.$$

Otherwise, we may estimate one of the θ_i's by $\frac{U_j c}{n}$ where $\theta_j = \theta_i$, but taking the intersection with all possible orders, we obtain

$$\mathbb{P}_\theta\{\widehat{\theta} \neq \theta\} \leqslant \sum_{j=1}^{c} \mathbb{P}_\theta\left(\left|\frac{U_j c}{n} - \theta_j\right| \geqslant \frac{1}{2}\right).$$

Conditionally on $W = \theta$, U_j has distribution $\mathscr{B}(n, \theta_j/c)$ and thus

$$\mathrm{Var}_\theta[U_j] = n\frac{\theta_j}{c}\left(1 - \frac{\theta_j}{c}\right) \leqslant n\frac{d}{c}.$$

By Bernstein's Inequality 3.9,

$$\mathbb{P}_\theta\left(\left|\frac{U_j}{n} - \frac{\theta_j}{c}\right| \geqslant \frac{1}{2c}\right) \leqslant 2\exp\left(-\frac{n/4c^2}{2\left(d/c + 1/6c\right)}\right) = 2\exp\left(-\frac{n}{8c(d + 1/6)}\right).$$

Hence,

$$\mathbb{P}(\widehat{\theta} \neq \theta) = \frac{1}{\Theta^{c,d}}\sum_{\theta \in \Theta^{c,d}} \mathbb{P}_\theta(\widehat{\theta} \neq \theta) \leqslant 2c\exp\left(-\frac{n}{8c(d + 1/3)}\right)$$

and we get

$$I(\Psi_{1:n}; W) \geqslant (\log_2 \mathrm{e})f(\lambda)\sqrt{c}\left(1 + o(1)\right)\left(1 - 2c\,\mathrm{e}^{-\frac{n}{8\lambda c^{3/2}}}\right).$$

Taking $c = (3n/16\lambda \log_2 n)^{2/3}$, we have

$$\overline{R}_n^\Psi(\Lambda) \geqslant \frac{f(\lambda)\log_2 \mathrm{e}}{\lambda^{1/3}}\left(\frac{3n}{16\log_2 n}\right)^{1/3}(1 + o(1)).$$

Looking at the table of f given in Dixmier and Nicolas [11, p. 151], this is optimized for $\lambda = 0.8$, which gives $f(\lambda) \approx 2.07236$.

3.6 Notes

A large part of this chapter comes from the article of Boucheron, Garivier and Gassiat [12], where one can find the description of a code, called the "censuring code", which takes advantage of the idea of truncation. This code is adapted in [4] (with an adaptive truncation) to make this code optimal (and adaptive) over envelop classes with exponential decay.

Acharya et al. [3] used a technique known as *Poissonization* (randomizing the number of observations according to a Poisson distribution with mean n) to obtain an upper bound scaling as $n^{1/\alpha}$ for the regret of the polynomial decay envelop class with parameter α, which matches the lower bound of Theorem 3.6.

Coding and data compression over large-size alphabets or infinite alphabets are largely open research fields. This chapter does not contain all known results on this subject, but gives a good survey. He and Yang [13] use the idea of grammar coding in the context of countable alphabets, whereas Foster, Stine and Wyner [14] study different variants of the redundancy and propose a code for memoryless sources with weakly decreasing distributions over the integers. Dhulipala and Orlitsky [15] estimate the minimax redundancy for Markov processes and hidden Markov chains.

The study of hidden Markov chains (investigated in the next chapter) is a very active research field, in finite or infinite alphabets. This is a classical model of communication channels. Numerous works are concerned with their entropy rate, its estimation, and its various properties.

References

1. P. Elias, Universal codeword sets and representations of the integers. IEEE Trans. Inf. Theory **21**, 194–203 (1975)
2. J. Kieffer, A unified approach to weak universal source coding. IEEE Trans. Inf. Theory **24**, 674–682 (1978)
3. J. Acharya, A. Jafarpour, A. Orlitsky, A.T. Suresh, Poissonization and universal compression of envelope classes, in *2014 IEEE International Symposium on Information Theory (ISIT)*, pp. 1872–1876 (IEEE, 2014)
4. D. Bontemps, Universal coding on infinite alphabets: exponentially decreasing envelopes. IEEE Trans. Inf. Theory **57**(3), 1466–1478 (2011). ISSN 0018-9448. http://dx.doi.org/10.1109/TIT. 2010.2103831
5. D. Haussler, M. Opper, Mutual information, metric entropy and cumulative relative entropy risk. Annals Stat. **25**, 2451–2492 (1997)
6. P. Massart, Concentration inequalities and model selection, in *Lecture Notes in Mathematics*. Lectures from the 33rd Summer School on Probability Theory held in Saint-Flour, July 6-23, 2003, With a foreword by Jean Picard, vol. 1896 (Springer, Berlin, 2007). ISBN 978-3-540-48497-4; 3-540-48497-3
7. G.M. Gemelos, T. Weissman, On the entropy rate of pattern processes. IEEE Trans. Inf. Theory **52**, 3994–4007 (2006)
8. A. Orlitsky, N.P. Santhanam, Speaking of infinity. IEEE Trans. Inf. Theory **50**, 2215–2230 (2004)
9. G. Hardy, S. Ramanujan, Asymptotic formulæ in combinatory analysis (Proc. London Math. Soc. **17**(2), 75–115, (1918)), in *Collected Papers of Srinivasa Ramanujan*, pp. 276–309 (AMS Chelsea Publ., Providence, RI, 2000)
10. A. Garivier, A lower bound for the maximin redundancy in pattern coding. Entropy **11**, 634–642 (2009)
11. J. Dixmier, J.-L. Nicolas, Partitions sans petits sommants, in *A Tribute to Paul Erdős*, pp. 121–152 (Cambridge Univ. Press, Cambridge, 1990)
12. S. Boucheron, A. Garivier, E. Gassiat, Coding on countably infinite alphabets. IEEE Trans. Inf. Theory **55**, 358–373 (2009)

13. D. He, E. Yang, The universality of grammar-based codes for sources with countably infinite alphabets. IEEE Trans. Inf. Theory **51**, 3753–3765 (2005)
14. D. Foster, R. Stine, A. Wyner, Universal codes for finite sequences of integers drawn from a monotone distribution. IEEE Trans. Inf. Theory **48**, 1713–1720 (2002)
15. A. Dhulipala, A. Orlitsky, Universal compression of Markov and related sources over arbitrary alphabets. IEEE Trans. Inf. Theory **53**, 4182–4190 (2006)

Chapter 4
Model Order Estimation

Abstract In the statistical approach known as model selection, a collection of models is given, and the statistical procedure for model selection stems from a compromise between model complexity and adequacy to the data. Different types of adequacy to data can be used, according to the question one is interested in. In some applications, the model order is a structural parameter with a precise interpretation regarding the phenomenon being studied. Here, we will be interested in the statistical estimation of a model's order, as well as its link with universal coding. We start by recalling the general framework of model selection, and briefly describe the MDL principle, as introduced by Rissanen, according to which "sparse coding" leads to a good compromise between complexity and adequacy. We will then consider more specifically the question of inferring the order of a model, focusing on two types of model collections: hidden Markov chains models, and population mixture models. We study in detail penalized maximum likelihood estimators. As it will be revealed, we will need to understand the likelihood's fluctuations, and we will see why this is a hard question for hidden Markov chains models and for population mixture models. We will see how results in universal coding allow for a first analysis of the likelihood's fluctuations and help to obtain consistency results, in particular for hidden Markov chains. However, this analysis is still sub-optimal, and the end of the chapter will be devoted to the study of independent random variables, in particular in population mixture models. In this situation, it is possible to carry out a precise study of the likelihood's fluctuations.
In each section, we will clarify in which set $\mathscr{X}$ the random variables take their values.

4.1 Generalities

4.1.1 Model Selection and the MDL Principle

Let $(X_n)_{n \in \mathbb{N}}$ be a sequence of random variables with distribution $\mathbb{P}$ and values in a complete separable metric space $\mathscr{X}$. Let $(\mathscr{M}_q)_{q \in \mathscr{Q}}$ be a given collection of models, i.e. for each $q \in \mathscr{Q}$, $\mathscr{M}_q$ is a class of distributions over $\mathscr{X}^{\mathbb{N}}$. For a given integer n, and

© Springer International Publishing AG, part of Springer Nature 2018   103
É. Gassiat, *Universal Coding and Order Identification by Model
Selection Methods*, Springer Monographs in Mathematics,
https://doi.org/10.1007/978-3-319-96262-7_4

for $X_1, \ldots, X_n$, we choose a model according to some specified statistical objective. The chosen model then allows us to perform statistical inference, estimation, and tests, in the framework given by the model. The model selection procedure is thus driven by the pursued statistical objective. This objective often reduces to a cost function measuring some gap between a characteristic of the (unknown) distribution $\mathbb{P}$ of the process $(X_n)_{n \in \mathbb{N}}$ and its estimator. This gap can often be decomposed in a term called "bias", measuring the quality of the approximation of $\mathbb{P}$ by the selected model, and a term called "variance", measuring estimation quality in the model. Model selection methods then often consist in a trade-off between bias and variance. This trade-off can be interpreted in terms of complexity: for instance, when each model $\mathcal{M}_q$ is parametrized by some finite-dimensional parameter $d(q)$, the larger $d(q)$, the higher the complexity. The bias term will then be small, but as the number of parameters to estimate increases, so does the variance. For instance, we have seen in previous chapters that if $\mathscr{X}$ is a finite set, the model collection $(\mathcal{M}_q)_{q \in \mathbb{N}}$, where $\mathcal{M}_q$ is the class of Markov chains' distributions of order q, is useful for approaching stationary ergodic source distributions, if we are interested in universal coding. We can also refine this model collection by considering classes of distributions of context tree sources. In this case, $\mathscr{Q}$ is the set of complete finite trees in $\mathscr{X}^*$ and $\mathcal{M}_q$ is the class of source distributions with context tree q.

We have seen, in the previous two chapters, that statistical ideas could help to interpret universal coding methods or understand compression capacities of some codes. In the 1970s, Rissanen [1] proposed a general model selection method according to the MDL principle (MDL for *Minimum Description Length*): choose the model which gives the shortest description of data, i.e. the shortest codeword length. The main idea is as follows. Given a sequence of models $(\mathcal{M}_q)_{q \in \mathscr{Q}}$, we design, for each $q \in \mathscr{Q}$, a coding distribution Q_q with good compression performance over $\mathcal{M}_q$ (in terms of regret or redundancy, for instance). To encode the data, we may proceed in the following way. We choose a coding distribution P over $\mathscr{Q}$, for instance the one given by Elias coding if $\mathscr{Q}$ is the set of the integers. We encode q according to P, and then the data according to Q_q. Note that the use of the expression "coding according to" implies that the coding method is known, and that the set must have been discretized to come down to a finite set over which arithmetic coding is used. A large part of the analysis of MDL methods is concerned with this discretization of the space and will not be discussed here. For our purpose, we will assume that the data coding has code length of order

$$- \log_2 P\,(q) - \log_2 Q_q\,(X_{1:n})\,.$$

Model selection by MDL then consists in choosing the model index $\widehat{q}_{\mathrm{MDL}}$ which minimizes this code length:

$$\widehat{q}_{\mathrm{MDL}} = \operatorname{argmin}_{q \in \mathscr{Q}} \left\{ - \log_2 P\,(q) - \log_2 Q_q\,(X_{1:n}) \right\}.$$

This general principle has several variants, see for instance the survey paper [2].

If Q_q is NML, then the order estimator is a penalized maximum likelihood estimator:

$$\widehat{q}_{\mathrm{MDL}} = \mathrm{argmax}_{q \in \mathcal{Q}} \Big\{ \sup_{Q \in \mathcal{M}_q} \log_2 Q(X_{1:n}) - R_n^*(\mathcal{M}_q) + \log_2 P(q) \Big\}.$$

Generally speaking, this leads us to consider penalized estimators of the form

$$\widehat{q}_n = \mathrm{argmax}_{q \in \mathcal{Q}} \big\{ \log_2 Q_q(X_{1:n}) - \mathrm{pen}(n, q) \big\}.$$

For a parametric collection of models, Rissanen's Theorem 2.13 suggests that the right order for $\mathrm{pen}(n, q)$, called the *penalty*, is $\log_2 n$ times half the number of parameters necessary to describe the model $\mathcal{M}_q$. This quantity is called the BIC penalty (for *Bayesian Information Criterion*).

There exist other model selection methods, and numerous works are concerned with evaluating these methods in terms of statistical performance according to the chosen criterion, see for instance [3]. In the remainder of the chapter, we will be specifically interested in the question of order identification as presented below, evaluated with the error probability, or with almost sure convergence (when the number of observations n tends to infinity). This framework is chosen when the order is a structural parameter which can be interpreted in terms of the phenomenon being studied and which is the interest parameter for the statistician.

4.1.2 Order Identification and Penalized Maximum Likelihood

We consider the case where $\mathcal{Q} = \mathbb{N}$ and where models are nested:

$$\mathcal{M}_q \subset \mathcal{M}_{q+1}, \quad q \in \mathbb{N}.$$

If $\mathcal{M} = \bigcup_{q \in \mathbb{N}} \mathcal{M}_q$ and $\mathbb{P} \in \mathcal{M}$, we then define the order of distribution $\mathbb{P}$ as:

$$q(\mathbb{P}) = \inf \big\{ q : \mathbb{P} \in \mathcal{M}_q \big\}.$$

Here are some classical examples:

- Order of a Markov chain: we say that $(X_n)_{n \in \mathbb{N}}$ is a *Markov chain of order q* if, for all integers $n \geqslant q + 1$, the conditional distribution of X_n given $X_{1:n-1}$ is equal to the conditional distribution of X_n given $X_{n-q:n-1}$. The model $\mathcal{M}_q$ is then defined as a set of distributions of Markov chains of order q.

- Order of a hidden Markov chain: we say that $(X_n)_{n\in\mathbb{N}}$ is a *hidden Markov chain* if there exists a finite set and a Markov chain $(Z_n)_{n\in\mathbb{N}}$ (of order 1) with values in that set such that, conditionally on $(Z_n)_{n\in\mathbb{N}}$, the random variables X_i are independent, and the conditional distribution of X_i only depends on the Markov chain through its value at time i. We call the size of the hidden states' space the *order of the hidden Markov chain*. The set $\mathcal{M}_q$ is then defined as the set of distributions of $(X_n)_{n\in\mathbb{N}}$ when the hidden states' space, denoted E_q, has q elements.
- Order of a population mixture: $(X_n)_{n\in\mathbb{N}}$ is a sequence of independent and identically distributed random variables, whose density is a convex combination of densities in a given family $\mathcal{G}$, which is generally assumed to be parametric. We say that the law is a *population mixture* and we call the number of elements in that convex combination the *order* (or number of populations). The set $\mathcal{M}_q$ is then defined as the set of distributions of sequences of i.i.d. random variables whose marginal density is a convex combination of q elements of $\mathcal{G}$.

In the first case, the order parameter characterizes the dependency structure of the sequence of random variables. In the two other cases, the order parameter characterizes a classification in different states, which can be modeled by a hidden variable, which may itself have a concrete interpretation. For instance, it may represent the state of a transmission channel for transmission of a data sequence in telecommunications, or the biological function of the considered region in a DNA sequence of codons.

We now give some simple heuristics about penalized maximum likelihood estimators. Let $(X_n)_{n\in\mathbb{N}}$ be a sequence of random variables with distribution $\mathbb{P}^\star \in \mathcal{M}$ such that $q(\mathbb{P}^\star) = q^\star$. In order to define the penalized maximum likelihood estimator, models $\mathcal{M}_q$ need to be dominated, and we will consider that for all q, the elements of $\mathcal{M}_q$ define, for all integers n, densities with respect to a dominating measure over $\mathcal{X}^n$. We then define

$$\widehat{q}_n = \mathrm{argmax}_{q\in\mathbb{N}}\Big\{ \sup_{Q\in\mathcal{M}_q} \log_2 Q\left(X_{1:n}\right) - \mathrm{pen}\left(n, q\right) \Big\}.$$

Since, for all integers q,

$$(\widehat{q}_n = q) \subset \Big(\sup_{Q\in\mathcal{M}_q} \log_2 Q\left(X_{1:n}\right) - \sup_{Q\in\mathcal{M}_{q^\star}} \log_2 Q\left(X_{1:n}\right) \geqslant \mathrm{pen}\left(n, q\right) - \mathrm{pen}(n, q^\star)\Big),$$

choosing the penalty $\mathrm{pen}\left(\cdot, \cdot\right)$ requires us to understand the behavior of the likelihood ratio process

$$\left(\Delta_n\left(q, q^\star\right)\right)_{q\in\mathbb{N}} = \Big\{ \sup_{Q\in\mathcal{M}_q} \log_2 Q\left(X_{1:n}\right) - \sup_{Q\in\mathcal{M}_{q^\star}} \log_2 Q\left(X_{1:n}\right) \Big\}_{q\in\mathbb{N}}. \qquad (4.1)$$

When $q < q^*$ and when the random variables are independent, for all $Q \in \mathcal{M}_q$,

$$\frac{1}{n} \left\{ \log_2 Q\left(X_{1:n}\right) - \log_2 \mathbb{P}^\star\left(X_{1:n}\right) \right\}$$

converges $\mathbb{P}^\star$-a.s. to minus the Kullback distance between Q and $\mathbb{P}^\star$, and under weak assumptions,

$$\frac{1}{n} \Big[\sup_{Q \in \mathcal{M}_q} \log_2 Q\left(X_{1:n}\right) - \sup_{Q \in \mathcal{M}_{q^*}} \log_2 Q\left(X_{1:n}\right) \Big]$$

converges to minus a generalized Kullback distance between $\mathbb{P}^\star$ and $\mathcal{M}_q$.

When $q \geqslant q^*$ and when $\mathcal{M}_q$ is a regular identifiable model,

$$2\Big\{ \sup_{Q \in \mathcal{M}_q} \log_2 Q\left(X_{1:n}\right) - \sup_{Q \in \mathcal{M}_{q^*}} \log_2 Q\left(X_{1:n}\right) \Big\}$$

converges in distribution to a chi-square whose number of degrees of freedom is the difference between the models' dimensions. In these cases, one may infer some simple conditions on penalty choices for which $\widehat{q}_n$ is consistent in probability, i.e. satisfies $\lim_{n \to +\infty} \mathbb{P}^\star\left(\widehat{q}_n \neq q^*\right) = 0$. Indeed, if the penalty satisfies that for all integers q, $\frac{1}{n}\mathrm{pen}(n, q)$ tends to 0 as n tends to infinity, and if, for $q < q^*$, the generalized Kullback distance between $\mathbb{P}^\star$ and $\mathcal{M}_q$ is strictly positive, then $\mathbb{P}^\star\left(\widehat{q}_n < q^*\right)$ tends to 0. If on the other hand, for all $q > q^*$, $\mathrm{pen}(n, q) - \mathrm{pen}(n, q^*)$ tends to $+\infty$, then for all $q > q^*$, $\mathbb{P}^\star\left(\widehat{q}_n = q\right)$ tends to 0. Hence, if an *a priori* bound on the order is known, $\widehat{q}_n$ is consistent in probability.

In contrast, these considerations are not enough to get consistency without any *a priori* bound on the order, and *a fortiori* to obtain almost sure consistency, i.e. that $\mathbb{P}^\star$-a.s., $\widehat{q}_n$ tends to q^* (so $\mathbb{P}^\star$-a.s. is equal to q^* for n large enough). In particular, in the case of hidden Markov chains and population mixtures, the likelihood ratio statistics $\Delta_n\left(q, q^*\right)$ when $q > q^*$ is no longer asymptotically distributed as half a chi-square.

To obtain consistency results $\mathbb{P}^\star$-a.s., we need to know how to evaluate the quantity $\mathbb{P}^\star\left(\widehat{q}_n = q\right)$, for all integers q and n. Evaluating $\mathbb{P}^\star\left(\widehat{q}_n = q\right)$, for all integers q and n, requires a thinner understanding of the trajectories in (4.1), quantified in terms of the order q and the number of observations n.

When the collection of models is that of Markov chains with finite state space, the maximum likelihood can be explicitly written, and a direct study is possible, see [4].

In the remainder of this chapter, we will first be interested in hidden Markov models. We will explain why the behavior of the likelihood ratio statistics cannot be understood in the classical framework. We will see how the inequalities obtained for coding with Krichevsky–Trofimov distributions can be used to evaluate likelihood ratios, by providing an upper bound $M(n, q) > 0$ for $\Delta_n\left(q, q^*\right)$ uniformly over $q > q^*$. Since

$$(\widehat{q}_n \leq q^\star) \supset \left(\forall q > q^\star, \ \Delta_n\left(q, q^\star\right) < \mathrm{pen}\,(n, q) - \mathrm{pen}\,\left(n, q^\star\right)\right),$$

it suffices to choose the penalty so that $\mathrm{pen}(n, q) \geq M(n, q) + \mathrm{pen}\,(n, q^\star)$ for all $q > q^\star$, which is guaranteed by the choice $\mathrm{pen}(n, q) = \sum_{k=1}^{q} M(n, q)$.

We will then be interest in population mixtures. In the case of sequences of independent variables, we will see that the behavior of the likelihood ratio can be written as the supremum of the square of an empirical process, and that a thin understanding of empirical processes can be used to show that $\log_2 \log_2 n$ is the minimal order of growth for the penalty and suffices to obtain an almost surely consistent estimator by penalized likelihood.

4.2   Hidden Markov Chains

Let us recall that the sequence of variables $(X_n)_{n \in \mathbb{N}}$ with values in $\mathscr{X}^{\mathbb{N}}$ is said to be a hidden Markov chain if there exists a sequence of variables $(Z_n)_{n \in \mathbb{N}}$, a Markov chain of order 1, such that, conditionally on $(Z_n)_{n \in \mathbb{N}}$, the sequence $(X_n)_{n \in \mathbb{N}}$ is a sequence of independent variables, and for all integers i, the conditional distribution of X_i given $(Z_n)_{n \in \mathbb{N}}$ only depends on Z_i. This conditional distribution is then called the emission distribution.

Here, we only consider Markov chains $(Z_n)_{n \in \mathbb{N}}$ with values in a finite set.

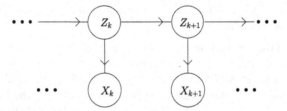

4.2.1   The Likelihood of Hidden Markov Chains

We will now try to understand the likelihood's behavior, and see that this behavior is not standard. Let $E_q = \{1, \ldots, q\}$ and $\mathscr{G} = \{g_\gamma, \gamma \in \Gamma\}$, with $\Gamma \subset \mathbb{R}^d$, be a set of densities over $\mathscr{X}$ with respect to a fixed dominant measure. The set $\mathscr{G}$ is the set of emission distributions. For all integers q, recall that $\mathscr{S}_q$ denotes the simplex of $\mathbb{R}^q$ and $\mathbb{S}_q$ denotes the set of parameters $(u_1, \ldots, u_{q-1}) \in \mathbb{R}^{q-1}$ such that, letting $u_q = 1 - u_1 - \cdots - u_{q-1}$, then $(u_1, \ldots, u_q) \in \mathscr{S}_q$. The distribution $\mathbb{P}_\theta$ of the hidden Markov chain $(X_n)_{n \in \mathbb{N}}$ is thus completely defined by $\theta = (\mu, \theta^1, \theta^2)$, where $\mu \in \mathbb{S}_q$ defines the initial distribution of the Markov chain $(Z_n)_{n \in \mathbb{N}}$, where $\theta^1 = (\theta_i^1)_{1 \leqslant i \leqslant q} \in \mathbb{S}_q^q$ defines the q transition probability distributions of the Markov chain $(Z_n)_{n \in \mathbb{N}}$, and where $\theta^2 = (\theta_i^2)_{1 \leqslant i \leqslant q} \in \Gamma^q$ defines the emission distributions' parameters. Denote by Θ_q the set of those parameters θ. The log-likelihood can then be written, for $\theta \in \Theta_q$, as

$$\ell_n(\theta) = \log_2 \Big[\sum_{z_{0:n} \in E_q^{n+1}} \mathbb{P}_\theta(Z_{0:n} = z_{0:n}) \prod_{i=1}^{n} g_{\gamma_{z_i}}(X_i) \Big],$$

with

$$\mathbb{P}_\theta(Z_{0:n} = z_{0:n}) = \mu(z_0) \prod_{i=0}^{n-1} (\theta^1_{z_i})_{z_{i+1}}.$$

This log-likelihood can be written in an additive way, by writing the joint density of $X_{1:n}$ as the product, for i going from 1 to n, of the conditional densities of X_i given $X_{1:i-1}$. We get

$$\ell_n(\theta) = \sum_{i=1}^{n} \log_2 \Big[\sum_{j=1}^{q} \mathbb{P}_\theta(Z_i = j \mid X_{1:i-1}) g_{\gamma_j}(X_i) \Big].$$

To understand the behavior of this additive quantity, we need to understand the behavior of $\mathbb{P}_\theta(Z_i = j \mid X_{1:i-1})$, called "prediction probabilities". In the book [5], several recent results obtained in this way are detailed. But these results do not apply to our situation. It turns out that if the order of the hidden Markov chain is $q^\star < q$, then the log-likelihood has an erratic behavior. This was observed by Gassiat and Kéribin [6] who showed that, for $q^\star = 1$ and $q = 2$, the likelihood ratio statistics

$$\sup_{\theta \in \Theta_2} \ell_n(\theta) - \sup_{\theta \in \Theta_1} \ell_n(\theta)$$

converges in probability to $+\infty$. We now explain this phenomenon.

Let $(X_n)_{n \in \mathbb{N}}$ be a sequence of independent and identically distributed random variables, each with density g_{γ_1} for some $\gamma_1 \in \Gamma$. A sub-model of the hidden Markov chains model of order 2 is given by the model of stationary hidden Markov chains of order 2, with fixed emission distribution parameters γ_1 and γ_2. In other words, letting

$$\begin{pmatrix} 1-a & a \\ b & 1-b \end{pmatrix}$$

be the transition matrix of the Markov chain $(Z_n)_{n \in \mathbb{N}}$ on 2 states, with $\mu(1) = \frac{b}{a+b}$ and $\mu(2) = \frac{a}{a+b}$ characterizing the initial stationary distribution of the Markov chain $(Z_n)_{n \in \mathbb{N}}$, the distribution of $(X_n)_{n \in \mathbb{N}}$ is obtained for $a = 0$, and the likelihood ratio statistics is lower-bounded by

$$\sup_{0 \leqslant a \leqslant 1, 0 \leqslant b \leqslant 1} \big[V_n(a, b) - V_n(0, b) \big],$$

where $V_n(a, b)$ is the log-likelihood of the considered sub-model. Letting $R_i(a, b)$ be the conditional probability, when the parameter is (a, b), that Z_i is equal to 2 given $X_{1:i-1}$, we have:

$$V_n(a, b) = \sum_{i=1}^{n} \log_2 \left[(1 - R_i(a, b)) g_{\gamma_1}(X_i) + R_i(a, b) g_{\gamma_2}(X_i) \right],$$

$$V_n(a, b) - V_n(0, b) = \sum_{i=1}^{n} \log_2 \left[(1 - R_i(a, b)) + R_i(a, b) \frac{g_{\gamma_2}}{g_{\gamma_1}}(X_i) \right].$$

The sequence of random variables $(R_i(a, b))_{i \geqslant 1}$ satisfies the induction relation

$$R_{i+1}(a, b) = \frac{a(1 - R_i(a, b)) g_{\gamma_1}(X_i) + (1 - b) R_i(a, b) g_{\gamma_2}(X_i)}{(1 - R_i(a, b)) g_{\gamma_1}(X_i) + R_i(a, b) g_{\gamma_2}(X_i)}.$$

In identifiable regular parametric models, the behavior of the gradient of the likelihood ratio monitors the asymptotic behavior. Here in contrast, the behavior of this gradient is not controlled. Indeed, since the parameter of the process $(X_n)_{n \in \mathbb{N}}$ can be $(0, b)$ for an arbitrary b in $[0, 1]$, we have to look at the derivative in a at point $(0, b)$ for arbitrary b in $[0, 1]$. This derivative is easily computed:

$$D_n(b) = \sum_{i=1}^{n} \left(\frac{g_{\gamma_2}}{g_{\gamma_1}}(X_i) - 1 \right) W_i(b),$$

where $W_i(b)$ is the derivative of $R_i(a, b)$ with respect to a at point $(0, b)$. By computation, the following induction relation follows:

$$W_{i+1}(b) = 1 + (1 - b) \frac{g_{\gamma_2}}{g_{\gamma_1}}(X_i) W_i(b), \quad W_1(b) = \frac{1}{b}.$$

For all i, the derivative $W_i(b)$ only depends on $X_{1:i-1}$ and is thus independent of X_i, so that $(W_i(b))_{i \geqslant 1}$ is a Markov chain. We can compute

$$\text{Var}[D_n(b)] = \text{Var}\left(\frac{g_{\gamma_2}}{g_{\gamma_1}}(X_1) \right) \sum_{i=1}^{n} E\left(W_i(b)^2 \right),$$

$$E\left(W_{i+1}(b)^2 \right) = \frac{2}{b} - 1 + (1 - b)^2 \left[\text{Var}\left(\frac{g_{\gamma_2}}{g_{\gamma_1}}(X_1) \right) + 1 \right] E\left(W_i(b)^2 \right),$$

and if b is such that

$$(1 - b)^2 \left[\text{Var}\left(\frac{g_{\gamma_2}}{g_{\gamma_1}}(X_1) \right) + 1 \right] > 1,$$

then $E(W_i(b)^2)$ grows exponentially fast to infinity. Hence, as n grows, if b is small enough, the derivative $D_n(b)$ has exponentially large fluctuations, and the likelihood ratio statistics is not tight. (The entire proof can be found in [6].)

We will now see that, even without completely understanding the behavior of the likelihood ratio statistics, the tools that have been developed in the previous chapters will help us to control the fluctuations of this statistics. The distribution of the hidden Markov source $(X_n)_{n \geqslant 1}$ will be denoted by $\mathbb{P}^\star$ and its order by $q^\star$.

4.2.2   Hidden Markov Chains with Values in a Finite Alphabet

We assume now that $\mathscr{X}$ is finite, and denote by m the number of elements of $\mathscr{X}$. The emission distributions are thus parametrized by elements of $\mathbb{S}_m$, and for all $q \in \mathbb{N}$, we may define

$$\Theta_q = \left\{ \theta = (\mu, \theta^1, \theta^2) : \ \mu \in \mathbb{S}_q, \ \theta^1 \in \mathbb{S}_q^q, \ \theta^2 \in \mathbb{S}_m^q \right\}$$

the set of parameters θ describing the distribution $\mathbb{P}_\theta$ of the hidden Markov chain. For all $x_{1:n} \in \mathscr{X}^n$, we have:

$$\mathbb{P}_\theta(x_{1:n}) = \sum_{z_{0:n} \in E_q^{n+1}} \mathbb{P}_\theta(z_{0:n}) \, \mathbb{P}_\theta(x_{1:n} \mid z_{1:n}) \, .$$

In order to control the maximum likelihood, we will define a mixture distribution (knwon as the Krichevsky–Trofimov distribution) and establish an inequality analogous to the inequalities obtained in Theorem 2.16, Propositions 2.18 and 2.19. This will provide an upper bound on the maximum likelihood ratio, and will help us obtain, by a change of distributions argument detailed below, an upper bound on the over-estimation probability, when estimating the order of a hidden Markov chain by penalized maximum likelihood.

We now endow Θ_q with a *prior* distribution ν such that μ is constant, equal to the uniform distribution over E_q, and the transition distributions θ_i^1, for $i = 1, \ldots, q$, are independent random variables with Dirichlet distribution of parameter $(\frac{1}{2}, \ldots, \frac{1}{2})$ over $\mathbb{S}_q$, independent of the emission distributions θ_i^2, which are independent random variables with Dirichlet distribution of parameter $(\frac{1}{2}, \ldots, \frac{1}{2})$ over $\mathbb{S}_m$.

We then define $\mathbb{KT}^q$ over $\mathscr{X}^{\mathbb{N}}$ by Kolmogorov's Extension Theorem from

$$\mathbb{KT}^q(x_{1:n}) = \sum_{z_{0:n} \in E_q^{n+1}} \int_{\Theta_q} \mathbb{P}_\theta(z_{0:n}) \, \mathbb{P}_\theta(x_{1:n} \mid z_{1:n}) \, \nu(\mathrm{d}\theta)$$

$$= \sum_{z_0 \in E_q} \frac{1}{q} \sum_{z_{1:n} \in E_q^n} \mathbb{KT}_{E_q}(z_{1:n} \mid z_0) \, \mathbb{KT}(x_{1:n} \mid z_{1:n})$$

for all integers n and for all $x_{1:n} \in \mathscr{X}^n$, where $\mathbb{KT}_{E_q}$ is the Krichevsky–Trofimov distribution given by the mixture of context tree sources with context tree the complete tree of depth 1 over E_q, and $\mathbb{KT}(. \,|\, z_{1:n})$ is the product of mixtures of memoryless distributions:

$$\mathbb{KT}(x_{1:n} \,|\, z_{1:n}) = \prod_{j \in E_q} \mathbb{KT}(x_{I_j}),$$

with $I_j = \{i : z_i = j\}$ (this set of indices depends on $z_{1:n}$) and $x_{I_j} = (x_i)_{i \in I_j}$. We then have the following result.

Proposition 4.1 *For all integers q and for all $x_{1:n} \in \mathscr{X}^n$,*

$$0 \leqslant \sup_{\theta \in \Theta_q} \log_2 \mathbb{P}_\theta(x_{1:n}) - \log_2 \mathbb{KT}^q(x_{1:n}) \leqslant \frac{1}{2} q(q + m - 2) \log_2 n + c(q, m)$$

with $c(q, m) = q\left[4 - \frac{1}{2}(q + m - 4)\log_2 q\right]$.

Proof (Proof of Proposition 4.1). We have

$$\sup_{\theta \in \Theta_q} \log_2 \frac{\mathbb{P}_\theta(x_{1:n})}{\mathbb{KT}^q(x_{1:n})} = \sup_{\theta \in \Theta_q} \log_2 \frac{\sum_{z_{0:n} \in E_q^{n+1}} \mathbb{P}_\theta(z_{0:n}) \, \mathbb{P}_\theta(x_{1:n} \,|\, z_{1:n})}{\sum_{z_0 \in E_q} \frac{1}{q} \sum_{z_{1:n} \in E_q^n} \mathbb{KT}_{E_q}(z_{1:n} \,|\, z_0) \, \mathbb{KT}(x_{1:n} \,|\, z_{1:n})}$$

$$\leqslant \log_2 q + \sup_{\theta \in \Theta_q} \log_2 \frac{\sum_{z_{0:n} \in E_q^{n+1}} \mathbb{P}_\theta(z_{1:n} \,|\, z_0) \, \mathbb{P}_\theta(x_{1:n} \,|\, z_{1:n})}{\sum_{z_0 \in E_q} \sum_{z_{1:n} \in E_q^n} \mathbb{KT}_{E_q}(z_{1:n} \,|\, z_0) \, \mathbb{KT}(x_{1:n} \,|\, z_{1:n})}$$

$$\leqslant \log_2 q + \sup_{\theta \in \Theta_q} \sup_{z_{0:n} \in E_q^{n+1}} \log_2 \frac{\mathbb{P}_\theta(z_{1:n} \,|\, z_0)}{\mathbb{KT}_{E_q}(z_{1:n} \,|\, z_0)} + \sup_{\theta \in \Theta_q} \sup_{z_{1:n} \in E_q^n} \log_2 \frac{\mathbb{P}_\theta(x_{1:n} \,|\, z_{1:n})}{\mathbb{KT}(x_{1:n} \,|\, z_{1:n})}.$$

On the other hand, by Proposition 2.18,

$$\sup_{\theta \in \Theta_q} \sup_{z_{0:n} \in E_q^{n+1}} \log_2 \frac{\mathbb{P}_\theta(z_{1:n} \,|\, z_0)}{\mathbb{KT}_{E_q}(z_{1:n} \,|\, z_0)} \leqslant \frac{q(q-1)}{2} \log_2 n - \frac{q(q-1)}{2} \log_2 q + 2q,$$

and by Theorem 2.16, for all $z_{1:n} \in E_q^n$,

$$\sup_{\theta \in \Theta_q} \log_2 \frac{\mathbb{P}_\theta(x_{1:n} \,|\, z_{1:n})}{\mathbb{KT}(x_{1:n} \,|\, z_{1:n})} \leqslant \sum_{j=1}^{q} \left(\frac{m-1}{2} \log_2 n_j + 2\right),$$

where $n_j = \sum_{i=1}^{n} 1_{z_i = j} = |I_j|$. Finally, by concavity and since $\sum_{j=1}^{q} n_j = n$,

$$\sum_{j=1}^{q} \log_2 n_j \leqslant q \log_2 \left(\frac{n}{q}\right). \qquad \square$$

Thanks to this inequality, the penalty in the penalized maximum likelihood estimator can be chosen such that this estimator converges almost surely, without any *a priori* bound. Let us recall that this estimator is defined by

$$\widehat{q}_n = \text{argmax}_{q \in \mathbb{N}} \Big\{ \sup_{\theta \in \Theta_q} \log_2 \mathbb{P}_\theta (X_{1:n}) - \text{pen}(n, q) \Big\}. \tag{4.2}$$

Theorem 4.2 *If for some $a > 2$, we have*

$$\text{pen}(n, q) = \sum_{k=1}^{q} \Big(\frac{k(k + m - 2) + a}{2} \Big) \log_2 n,$$

then $\mathbb{P}^\star$ a.s., for n large enough, $\widehat{q}_n \leqslant q^\star$.

Proof (Proof of Theorem 4.2). The proof of the theorem boils down to using the inequality of Proposition 4.1 in order to obtain, by change of distributions, an upper bound on the over-estimation probability, and concluding with the Borel–Cantelli Lemma. For all integers $q > q^\star$,

$$\mathbb{P}^\star \big(\widehat{q}_n = q \big) \leqslant \mathbb{P}^\star \Big(\sup_{\theta \in \Theta_q} \log_2 \mathbb{P}_\theta(X_{1:n}) - \text{pen}(n, q) \geqslant \log_2 \mathbb{P}^\star (X_{1:n}) - \text{pen}\big(n, q^\star\big) \Big)$$

and, by Proposition 4.1,

$$\mathbb{P}^\star \big(\widehat{q}_n = q \big) \leqslant \mathbb{P}^\star \Big\{ \log_2 \mathbb{KT}^q (X_{1:n}) + \frac{1}{2} q(q + m - 2) \log_2 n + c(q, m)$$
$$- \text{pen}(n, q) + \text{pen}(n, q^\star) \geqslant \log_2 \mathbb{P}^\star (X_{1:n}) \Big\}.$$

Let $A_{n,q}$ be the set of $x_{1:n} \in \mathcal{X}^n$ such that

$$\log_2 \mathbb{KT}^q (x_{1:n}) + \frac{1}{2} q(q + m - 2) \log_2 n + c(q, m) - \text{pen}(n, q) + \text{pen}\big(n, q^\star\big)$$

is larger or equal to $\log_2 \mathbb{P}^\star(x_{1:n})$. If $x_{1:n}$ belongs to $A_{n,q}$, then

$$\mathbb{P}^\star (x_{1:n}) \leqslant \mathbb{KT}^q (x_{1:n}) \, 2^{\frac{1}{2} q(q+m-2) \log_2 n + c(q,m) - \text{pen}(n,q) + \text{pen}(n,q^\star)}.$$

The key-argument is to use the inequality and a change of distributions in such a way that

$$\mathbb{P}^\star(\widehat{q}_n = q) \leqslant \sum_{x_{1:n} \in A_{n,q}} \mathbb{KT}^q (x_{1:n}) \, 2^{\frac{1}{2} q(q+m-2) \log_2 n + c(q,m) - \text{pen}(n,q) + \text{pen}(n,q^\star)}$$
$$\leqslant 2^{\frac{1}{2} q(q+m-2) \log_2 n + c(q,m) - \text{pen}(n,q) + \text{pen}(n,q^\star)},$$

and to choose the penalty in such a way that the sum, over q and n, of the above quantity is finite.

Indeed, with the penalty given in the statement of Theorem 4.2, we have, as soon as $q > q^\star$,

$$\frac{1}{2}q(q + m - 2)\log_2 n + c\,(q, m) - \text{pen}\,(n, q) + \text{pen}\,(n, q^\star)$$

$$= \left[\frac{1}{2}q(q + m - 2) - \sum_{k=q^\star+1}^{q} \frac{k(k + m - 2) + a}{2}\right]\log_2 n + c\,(q)$$

$$\leqslant -\frac{1}{2}a(q - q^\star)\log_2 n + c\,(q, m)$$

and since the function $q \mapsto c\,(q, m)$ is bounded over $\mathbb{N}$, for some constant $C > 0$:

$$\mathbb{P}^\star(\widehat{q}_n > q^\star) \leqslant \sum_{q > q^\star} n^{-\frac{1}{2}a(q-q^\star)}2^{c(q,m)} \leqslant Cn^{-\frac{1}{2}a}.$$

Hence, for $a > 2$, we have $\sum_{n \in \mathbb{N}} \mathbb{P}^\star(\widehat{q}_n > q^\star) < +\infty$, and the proof is complete thanks to the Borel–Cantelli Lemma.

To obtain the almost sure consistency of the estimator, we need to prove a Shannon–Breiman–McMillan Theorem for likelihood ratios of hidden Markov chains with values in a finite alphabet, and then that the difference between the maximum log-likelihood in a model of order $q < q^\star$ and the maximum log-likelihood in a model of order $q^\star$, normalized by n, converges to minus the infimum of the divergence rate previously obtained for those two models, and finally that this quantity is strictly negative. This proof can be found in [7].

4.2.3 Hidden Markov Chains with Gaussian Emission

The inequalities obtained previously with Krichevsky–Trofimov distributions hold for any trajectory, but heavily rely on the fact that $\mathscr{X}$ is a finite set. We now wish to use analogous tools in the case where random variables take real values, for Gaussian emission distributions. As we will see, this is possible, but it introduces in the upper bound a random term that needs to be controlled.

We now let

$$\mathscr{X} = \mathbb{R}.$$

Emission distributions are Gaussian distributions with equal known variance, i.e. for all integers n, the conditional distribution of X_n given $Z_n = i$ is the Gaussian distribution $\mathscr{N}(m_i, \sigma^2)$, where $\sigma > 0$ is known. This time, we have

$$\Theta_q = \big\{\theta = (\mu, \theta^1, \theta^2) : \mu \in \mathbb{S}_q, \ \theta^1 \in \mathscr{S}_q^q, \ \theta^2 \in \mathbb{R}^q\big\}.$$

The finite-dimensional marginals are absolutely continuous with respect to Lebesgue measure, and if we denote by g_θ the n-dimensional marginal density of $\mathbb{P}_\theta$ (n is omitted for ease of notation), we have, for all $x_{1:n} \in \mathbb{R}^n$, and with the previous notation,

$$g_\theta(x_{1:n}) = \sum_{z_{0:n} \in E_q^{n+1}} \mathbb{P}_\theta(z_{0:n}) \prod_{j=1}^q \prod_{i \in I_j} \phi_\sigma(x_i - m_j),$$

where ϕ_σ is the density of the normal distribution $\mathscr{N}(0, \sigma^2)$.

We endow Θ_q with a *prior* distribution ν such that μ is constant, equal to the uniform distribution over E_q, and the transition distributions θ_i^1, for $i = 1, \ldots, q$, are independent random variables with Dirichlet distribution of parameter $(\frac{1}{2}, \ldots, \frac{1}{2})$ over $\mathbb{S}_q$, independent of θ_i^2, which are independent random variables with Gaussian distribution $\mathscr{N}(0, \tau^2)$, for $\tau > 0$.

We then define a probability Q_q over $\mathbb{R}^n$ with density with respect to the Lebesgue measure

$$p_q(x_{1:n}) = \int_{\Theta_q} g_\theta(x_{1:n}) \nu(\mathrm{d}\theta)$$

$$= \sum_{z_0 \in E_q} \frac{1}{q} \sum_{z_{1:n} \in E_q^n} \mathbb{KT}_{E_q}(z_{1:n} \mid z_0) \prod_{j=1}^q \int \Big[\prod_{i \in I_j} \phi_\sigma(x_i - m_j) \Big] \phi_\tau(m_j) \mathrm{d} m_j,$$

which, letting $\bar{x}_j = \frac{1}{n_j} \sum_{i \in I_j} x_i$, $p_q(x_{1:n})$, is equal to

$$\sum_{z_0 \in E_q} \frac{1}{q} \sum_{z_{1:n} \in E_q^n} \mathbb{KT}_{E_q}(z_{1:n} \mid z_0) \frac{1}{(\sigma\sqrt{2\pi})^n} \prod_{j=1}^q \frac{1}{(1 + n_j \tau^2/\sigma^2)^{\frac{1}{2}}} \Upsilon$$

with

$$\Upsilon = \exp\left\{ -\frac{\sum_{i \in I_j} x_i^2}{2\sigma^2} + \frac{n_j^2 \tau^2}{2\sigma^4(1 + n_j \tau^2/\sigma^2)} \bar{x}_j^2 \right\}.$$

For all $x_{1:n} \in \mathscr{X}^n$, we let $(|x|_{(k)})_{1 \leqslant k \leqslant n}$ be the n-tuple given by the increasing re-ordering of $(|x|_k)_{1 \leqslant k \leqslant n}$. Likewise, $(|X|_{(k)})_{1 \leqslant k \leqslant n}$ is the order statistics of $(|X|_k)_{1 \leqslant k \leqslant n}$. We then have the following result.

Proposition 4.3 *For all integers q and for all $x_{1:n} \in \mathscr{X}^n$,*

$$0 \leqslant \sup_{\theta \in \Theta_q} \log_2 g_\theta(x_{1:n}) - \log_2 p_q(x_{1:n}) \leqslant \frac{q^2}{2} \log_2 n + \frac{q}{2\tau^2} |x|_{(n)}^2 + d(q, \tau)$$

with $d(q, \tau) = 2q - \frac{1}{2} q(q-3) \log_2 q + \frac{\tau^2}{2\sigma^2}$.

Proof (Proof of Proposition 4.3). We have

$$\sup_{\theta \in \Theta_q} \prod_{j=1}^{q} \prod_{i \in I_j} \phi_\sigma \left(x_i - m_j\right) = \prod_{j=1}^{q} \prod_{i \in I_j} \phi_\sigma \left(x_i - \bar{x}_j\right)$$

$$= \frac{1}{(\sigma\sqrt{2\pi})^n} \prod_{j=1}^{q} \exp\left\{ -\frac{\sum_{i \in I_j} x_i^2}{2\sigma^2} + \frac{n_j \bar{x}_j^2}{2\sigma^2} \right\}$$

and we proceed as for Proposition 4.1, so that

$$\sup_{\theta \in \Theta_q} \log_2 \frac{g_\theta \left(x_{1:n}\right)}{q_q \left(x_{1:n}\right)} \leqslant \log_2 q + \frac{q(q-1)}{2} \log_2 n - \frac{q(q-1)}{2} \log_2 q + 2q$$

$$+ \sup_{z_{1:n} \in E_q^n} \left\{ \frac{1}{2} \sum_{j=1}^{q} \log_2 \left(1 + \frac{n_j \tau^2}{\sigma^2}\right) + \sum_{j=1}^{q} \frac{n_j}{2\sigma^2(1 + n_j \tau^2/\sigma^2)} \bar{x}_j^2 \right\}.$$

On the other hand, by concavity of the logarithm function,

$$\frac{1}{2} \sum_{j=1}^{q} \log_2 \left(1 + \frac{n_j \tau^2}{\sigma^2}\right) \leqslant \frac{q}{2} \log_2 \left(1 + \frac{n\tau^2}{q\sigma^2}\right) = \frac{q}{2} \log_2 n + \frac{q}{2} \log_2 \left(\frac{1}{n} + \frac{\tau^2}{q\sigma^2}\right)$$

$$\leqslant \frac{q}{2} \log_2 n + \frac{q}{2} \log_2 \left(1 + \frac{\tau^2}{q\sigma^2}\right) \leqslant \frac{q}{2} \log_2 n + \frac{\tau^2}{2\sigma^2}.$$

Also,

$$\frac{n_j}{2\sigma^2(1 + \frac{n_j \tau^2}{\sigma^2})} \leqslant \frac{1}{2\tau^2}, \quad \sum_{j=1}^{q} \bar{x}_j^2 \leqslant q|x|_{(n)}^2. \qquad \square$$

To calibrate the penalty in the estimator (4.2), we thus have to control the random variable $|X|_{(n)}^2$. The following theorem gives an adequate calibration for the penalty.

Theorem 4.4 *If, for some a > 2,*

$$\text{pen}\,(n, q) = \sum_{k=1}^{q} \frac{1}{2}(k^2 + a) \log_2 n \,,$$

then $\mathbb{P}^\star$-a.s., for n large enough, $\widehat{q}_n \leqslant q^\star$.

Proof (Proof of Theorem 4.4). The proof works as in the case of finite emissions (use of the inequality in Proposition 4.3 and change of distributions), but with a random upper bound that needs to be calibrated. We will make use of the parameter of the mixture distribution Q_q as an additional tool. If $(t_n)_{n \in \mathbb{N}}$ is a sequence of non-negative real numbers, we first write

$$\mathbb{P}^*\left(\widehat{q}_n > q^*\right) = \mathbb{P}^*\left(\widehat{q}_n > q^* \text{ and } |X|^2_{(n)} \leqslant t_n\right) + \mathbb{P}^*\left(\widehat{q}_n > q^* \text{ and } |X|^2_{(n)} \geqslant t_n\right)$$

$$\leqslant \mathbb{P}^*\left(\widehat{q}_n > q^* \text{ and } |X|^2_{(n)} \leqslant t_n\right) + \mathbb{P}^*\left(|X|^2_{(n)} \geqslant t_n\right)$$

$$= \mathbb{P}^*\left(|X|^2_{(n)} \geqslant t_n\right) + \sum_{q > q^*} \mathbb{P}^*\left(\widehat{q}_n = q \text{ and } |X|^2_{(n)} \leqslant t_n\right).$$

Now, appealing to Proposition 4.3 and denoting by θ^* the parameter of Θ_{q^*} characterizing $\mathbb{P}^*$, for q and n fixed:

$$\mathbb{P}^*\left(\widehat{q}_n = q \text{ and } |X|^2_{(n)} \leqslant t_n\right)$$

$$\leqslant \mathbb{P}^*\left(\log_2 p_q(X_{1:n}) + \frac{1}{2}q^2 \log_2 n + d(q, \tau) + \frac{q}{2\tau^2}|X|^2_{(n)}\right.$$

$$\left. -\mathrm{pen}(n, q) + \mathrm{pen}(n, q^*) \geqslant \log_2 g_{\theta^*}(X_{1:n}) \text{ and } |X|^2_{(n)} \leqslant t_n\right)$$

$$\leqslant \mathbb{P}^*\left(\log_2 p_q(X_{1:n}) + \frac{1}{2}q^2 \log_2 n + d(q, \tau) + \frac{q}{2\tau^2}t_n\right.$$

$$\left. -\mathrm{pen}(n, q) + \mathrm{pen}(n, q^*) \geqslant \log_2 g_{\theta^*}(X_{1:n})\right).$$

We proceed with the change of distributions, so that if $A_{n,q} \subset \mathbb{R}^n$ is defined by

$$A_{n,q} = \left\{x_{1:n} : \log_2 p_q(x_{1:n}) + \frac{1}{2}q^2 \log_2 n + d(q, \tau) + \frac{q}{2\tau^2}t_n\right.$$

$$\left. - \mathrm{pen}(n, q) + \mathrm{pen}(n, q^*) \geqslant \log_2 g_{\theta^*}(x_{1:n})\right\},$$

we have

$$\mathbb{P}^*\left(\widehat{q}_n = q \text{ and } |X|^2_{(n)} \leqslant t_n\right)$$

$$\leqslant \int_{x_{1:n} \in A_{n,q}} g_{\theta^*}(x_{1:n})\, \mathrm{d}x_1 \ldots \mathrm{d}x_n$$

$$\leqslant 2^{\frac{1}{2}q^2 \log_2 n + d(q,\tau) + \frac{q}{2\tau^2}t_n - \mathrm{pen}(n,q) + \mathrm{pen}(n,q^*)} \int_{x_{1:n} \in A_{n,q}} p_q(x_{1:n})\, \mathrm{d}x_1 \ldots \mathrm{d}x_n$$

$$\leqslant 2^{\frac{1}{2}q^2 \log_2 n + d(q,\tau) + \frac{q}{2\tau^2}t_n - \mathrm{pen}(n,q) + \mathrm{pen}(n,q^*)}.$$

Now, we choose $t_n = 5\sigma^2 \log_2 n$ and $\tau^2 = \sigma^2 \sqrt{\log_2 n}$. We then have, n being fixed, and for all $q > q^*$:

$$\frac{1}{2}q^2 \log_2 n + d(q, \tau) + \frac{q}{2\tau^2}t_n - \mathrm{pen}(n, q) + \mathrm{pen}(n, q^*)$$

$$\leqslant -\frac{1}{2}a(q - q^*)\log_2 n + \frac{1}{2}(5q + 1)\sqrt{\log_2 n} + 2q - \frac{1}{2}q(q - 3)\log_2 q$$

$$-\left(\frac{a}{2} - \frac{5}{2\sqrt{\log_2 n}}\right)(q - q^*)\log_2 n + \frac{5q^* + 1}{2}\sqrt{\log_2 n} + {+}2q - \frac{q(q - 3)}{2}\log_2 q$$

and since the function $q \mapsto 2q - \frac{1}{2}q(q-3)\log_2 q$ is bounded over $\mathbb{N}$, for some constant $C > 0$:

$$\mathbb{P}^\star\left(\widehat{q}_n > q^\star \text{ and } |X|^2_{(n)} \leqslant t_n\right) \leqslant \left(\frac{a}{2} - \frac{5\sigma^2}{2\tau^2}\right)(q - q^\star)2^{d(q) + \frac{5\sigma^2 q^\star}{2\tau^2}} \leqslant Cn^{-\left(\frac{a}{2} - \frac{5(q^\star+2)}{\sqrt{\log_2 n}}\right)}.$$

This entails that if $a > 2$, then

$$\sum_{n \in \mathbb{N}} \mathbb{P}^\star\left(\widehat{q}_n > q^\star \text{ and } |X|^2_{(n)} \leqslant t_n\right) < +\infty.$$

Now,

$$\mathbb{P}^\star\left(|X|^2_{(n)} \geqslant t_n\right) = 1 - E\left[\prod_{i=1}^n \mathbb{P}^\star\left(|X_i|^2 \leqslant t_n \mid Z_i\right)\right] \leqslant 1 - \left[\mathbb{P}\left(|U| \leqslant \frac{\sqrt{t_n} + M}{\sigma}\right)\right]^n,$$

where $M = \max_{i=1,\dots,q} |m_i^\star|$ and U is distributed as $\mathcal{N}(0,1)$. It follows that for n large enough, and for some constant C,

$$\mathbb{P}^\star\left(|X|^2_{(n)} \geqslant 5\sigma^2 \log_2 n\right) \leqslant \frac{C}{n^{3/2}}.$$

The proof is concluded by appealing to the Borel–Cantelli Lemma.

To obtain the almost sure consistency of the estimator, we need to prove a Shannon–Breiman–McMillan Theorem for likelihood ratios of hidden Markov chains with Gaussian emission, and then that the difference between the maximum log-likelihood in a model of order $q < q^\star$ and the maximum log-likelihood in a model of order $q^\star$, normalized by n, converges to minus the infimum of the divergence rate previously obtained for those two models, and finally that this quantity is strictly negative. This proof can be found in [8].

Remark 4.1 To calibrate penalties, we used, in both cases, the method described in Sect. 4.1.2. We choose $\mathrm{pen}(n, q) = \sum_{k=1}^q \widetilde{M}(n, q)$ so as to obtain $\mathrm{pen}(n, q) \geq M(n, q) + \mathrm{pen}(n, q^\star)$ for all $q > q^\star$, where $M(n, q) > 0$ is an upper bound on $\Delta_n(q, q^\star)$ uniformly in $q > q^\star$. Note that the main term in $M(n, q)$ (and thus in $\widetilde{M}(n, q)$) is $\log_2 n$ times half the effective dimension of the parameter space (i.e. without accounting for the initial distribution of the Markov chain), namely $q(q + m - 2)$ for hidden Markov chains with values in a finite alphabet, and $q(q-1) + q = q^2$ for hidden Markov chains with Gaussian emission.

4.3   Independent Variables and Population Mixtures

We assume here that $\mathscr{X}$ is a complete metric space (endowed with its Borel sigma-field) and that $(X_n)_{n \in \mathbb{N}}$ is a sequence of independent random variables with distribution $\mathbb{P}^\star$ over $\mathscr{X}^{\mathbb{N}}$. Let μ be a non-negative measure over $\mathscr{X}$. We assume that the first marginal of $\mathbb{P}^\star$ (the probability distribution of each X_n) has density $f^\star$ with respect to μ. We then denote by $\mathscr{M}_q$ a (weakly) increasing sequence of sets of densities over $\mathscr{X}$ with respect to μ. Let $\mathscr{M} = \bigcup_{q \in \mathbb{N}} \mathscr{M}_q$ and $q^\star$ be the order of $\mathbb{P}^\star$, i.e.

$$q^\star = \inf \left\{ q : f^\star \in \mathscr{M}_q \right\}.$$

We are particularly interested in the case of population mixtures. Let

$$\mathscr{G} = \left\{ g_\gamma, \ \gamma \in \Gamma \right\},$$

with $\Gamma \subset \mathbb{R}^d$, be a family of probability densities with respect to μ over $\mathscr{X}$. We define, for all integers q,

$$\mathscr{M}_q = \left\{ \sum_{i=1}^{q} \pi_i g_{\gamma_i}, \ 0 \leqslant \pi_i, \ \gamma_i \in \Gamma, i = 1, \ldots, q, \sum_{i=1}^{q} \pi_i = 1 \right\}. \qquad (4.3)$$

We denote by $\ell_n(f)$ the log-likelihood, for a probability density f with respect to μ over $\mathscr{X}$:

$$\ell_n(f) = \sum_{i=1}^{n} \log f(X_i).$$

For q fixed, the model $\mathscr{M}_q$ is parametric. If $q > q^\star$, we have $f^\star \in \mathscr{M}_q$. However, classical results in parametric asymptotic statistics do not apply. Indeed, the model $\mathscr{M}_q$ is not identifiable. If we write

$$f^\star = \sum_{i=1}^{q^\star} \pi_i^\star g_{\gamma_i^\star},$$

with $\pi_i^\star > 0, \gamma_i^\star \neq \gamma_j^\star$ for $i \neq j$ and $i, j = 1, \ldots, q^\star$, and if

$$f = \sum_{i=1}^{q} \pi_i g_{\gamma_i},$$

then we have $f = f^\star$ as soon as $\pi_i = \pi_i^\star$ and $\gamma_i = \gamma_i^\star$ for $i = 1, \ldots, q^\star$, and $\pi_i = 0$ and γ_i arbitrary for $i = q^\star + 1, \ldots, q$, or as soon as $\pi_i = \pi_i^\star$ and $\gamma_i = \gamma_i^\star$ for $i = 1, \ldots, q^\star - 1$, and $\sum_{i=q^\star}^{q} \pi_i = \pi_{q^\star}^\star$ and $\gamma_i = \gamma_{q^\star}^\star$ for $i = q^\star, \ldots, q$, for instance.

The likelihood ratio statistics $\sup_{f \in \mathcal{M}_q} \ell_n(f) - \sup_{f \in \mathcal{M}_{q^*}} \ell_n(f)$ does not converge in distribution to half a chi-square.

In this section, we give an analysis of the likelihood ratio statistics, whose purpose is to help understand its fluctuations, so as to be able to describe its asymptotic behavior, in distribution and almost surely. This analysis is conducted in a general way in the context of a weakly increasing sequence of models, and is applied to population mixture models. The basic idea is to obtain an approximation of the likelihood ratio statistics by an empirical process, and to make use of modern tools pertaining to the theory of empirical processes. This idea stems from the following heuristics.

If, for $\delta > 0$, $(f_t)_{t \in [0,\delta]}$ is a regular sub-model of $\mathcal{M}_q$ such that $f_0 = f^*$ and with score function $\dot{s}_0$, we have, under some regularity and domination assumptions (see [9] for definitions and results):

$$
\sup_{t \in (0,\delta]} \ell_n(f_t) - \ell_n(f^*) = \frac{1}{2\|\dot{s}_0\|^2} \left(\left[\frac{1}{\sqrt{n}} \sum_{i=1}^n \dot{s}_0(X_i) \right] \vee 0 \right)^2 + o_{\mathbb{P}^*}(1),
$$

where $\| \cdot \|$ is the norm of $L^2(f^* \mathrm{d}\mu)$.

Denote by ν_n the empirical process, i.e., if g is a function of $L^2(f^* \mathrm{d}\mu)$,

$$
\nu_n(g) = \frac{1}{\sqrt{n}} \sum_{i=1}^n \big(g(X_i) - E[g(X_i)] \big).
$$

If $d_0 = \dot{s}_0 / \|\dot{s}_0\|$ is the normalized score, then

$$
\sup_{t \in (0,\delta]} \ell_n(f_t) - \ell_n(f^*) = \frac{1}{2} \big(\nu_n(d_0) \vee 0 \big)^2 + o_{\mathbb{P}^*}(1).
$$

Our goal here is to find appropriate conditions (not too strong) under which the supremum can be taken over a rich enough class of sub-models and to obtain in this way an approximation of the likelihood ratio by half the supremum of the squared positive part of an empirical process, the supremum being taken over a set of normalized scores. This would give the asymptotic distribution of the likelihood ratio statistics, under $\mathbb{P}^*$, and under contiguous alternatives. If this approximation holds almost surely, the functional law of the iterated logarithm states that the order of growth of the likelihood ratio is $\log \log n$. To calibrate the penalty in an order estimation procedure based on penalized maximum likelihood, one needs to obtain a uniform control in the order q, simultaneously for all models. As we will see, this will help us to obtain that $\log \log n$ times an increasing function of the order is indeed the minimum penalty and that this minimum penalty is sufficient, up to constant factors, to make the penalized maximum likelihood order estimator almost surely consistent, without any *prior* bound.

4.3.1 Some Tools

Let $\mathscr{D}$ be a subset of $L^2(f^\star d\mu)$. We will be interested in the process $(\nu_n(d))_{d\in\mathscr{D}}$, which is a real-valued stochastic process over $\mathscr{D}$, regarded as an element of $L^\infty(\mathscr{D})$ (when also $\mathscr{D} \subset L^\infty(\mathscr{D})$). Our goal is to understand asymptotic behaviors, in distribution and almost surely. To obtain convergence in distribution, we need to:

- check convergence in distribution of the finite-dimensional marginals, i.e. of the vectors $(\nu_n(d_1), \ldots, \nu_n(d_m))$ for all integers m, and all $d_1, \ldots, d_m$ in $\mathscr{D}$, which follows from the Central Limit Theorem,
- check tightness of the process in $L^\infty(\mathscr{D})$.

To obtain almost sure results, we can use the Borel–Cantelli Lemma. This assumes that we have deviation inequalities for the empirical process. Such results require conditions on the complexity of the class over which the empirical process is considered. We will mainly use results that are obtained under conditions known as bracketing entropy conditions, which we now define.

If ℓ and u are two functions in $L^2(f^\star d\mu)$, the bracket $[\ell, u]$ is the set of functions g in $L^2(f^\star d\mu)$ such that for all $x \in \mathscr{X}, \ell(x) \leqslant g(x) \leqslant u(x)$. We say that the bracket has *size* δ if $\|\ell - u\| \leqslant \delta$. For all $\delta > 0$, denote by $N(\mathscr{D}, \delta)$ the minimum number of brackets of size δ that is needed to cover $\mathscr{D}$. The bracketing entropy is then given by $\log N(\mathscr{D}, \delta)$. This is a weakly decreasing function of δ. The fact that the integral $\int_0^1 \sqrt{\log N(\mathscr{D}, u)} du$ is finite implies that $\log N(\mathscr{D}, u)$ is finite for all u, and that the set $\mathscr{D}$ has an envelop function in $L^2(f^\star d\mu)$, i.e. a function $D \in L^2(f^\star d\mu)$ such that for all $d \in \mathscr{D}, |d| \leqslant D$. Then, $\mathscr{D}$ is relatively compact in $L^2(f^\star d\mu)$. Moreover, $|\nu_n(d)|$ is uniformly upper-bounded, for $d \in \mathscr{D}$, by $\frac{2}{\sqrt{n}} \sum_{i=1}^n (D(X_i) + ED(X_i))$, so that $(\nu_n(d))_{d\in\mathscr{D}} \in L^\infty(\mathscr{D})$.

If we define $(W(d))_{d\in\mathscr{D}}$ as the centered Gaussian process over $\mathscr{D}$ with covariance function $\langle \cdot, cdot \rangle$ for the scalar product of $L^2(f^\star d\mu)$, condition $\int_0^1 \sqrt{\log N(\mathscr{D}, u)} du < +\infty$ implies continuity of the Gaussian process, and convergence, in $L^\infty(\mathscr{D})$, of the empirical process $(\nu_n(d))_{d\in\mathscr{D}}$ to the Gaussian process $(W(d))_{d\in\mathscr{D}}$. On the other hand, we can construct a probability space over which there exist versions of $(\nu_n(d))_{d\in\mathscr{D}}$ and $(W(d))_{d\in\mathscr{D}}$ such that the approximation holds almost surely. We then have an inequality, known as a *maximal inequality*:

$$E\Big| \sup_{d\in\mathscr{D}} \nu_n(d) \Big| \leqslant \int_0^{\|D\|} \sqrt{\log N(\mathscr{D}, u)} du. \tag{4.4}$$

For more on these notions as well as on their consequences in statistics, see for instance [10].

Still under the assumption that $\int_0^1 \sqrt{\log N(\mathscr{D}, u)} du < +\infty$, the empirical process over $\mathscr{D}$ satisfies a functional law of the iterated logarithm, see for instance [11]. More precisely, under this assumption, $\mathbb{P}^\star$-a.s., the sequence

$$\left[\left(\frac{\nu_n(d)}{\sqrt{2\log\log n}}\right)_{d\in\mathscr{D}}\right]_{n\in\mathbb{N}}$$

is relatively compact in $L^\infty(\mathscr{D})$, and the set of its accumulation points coincides with

$$\{d \mapsto \langle d, g \rangle : g \in L_0^2(f^\star d\mu)\},$$

where

$$L_0^2(f^\star d\mu) = \left\{g \in L^2(f^\star d\mu) : \int gf^\star d\mu = 0, \int g^2 f^\star d\mu \leqslant 1\right\}.$$

4.3.2   Remarks on Penalty Devising

Assumptions guaranteeing that $\mathbb{P}^\star$ a.s., for n large enough, $\widehat{q}_n \geqslant q^\star$ are classical. For instance, it suffices to assume that the set of log-densities in models with order less than or equal to $q^\star$ is $\mathbb{P}^\star$-Glivenko–Cantelli, and that the infimum of the Kullback distance between $f^\star$ and f, for $f \in \bigcup_{q<q^\star} \mathscr{M}_q$, is strictly positive. Then, if the penalty satisfies

$$\forall q, \quad \lim_{n\to+\infty} \frac{\text{pen}(n,q)}{n} = 0,$$

we have $\widehat{q}_n \geqslant q^\star$, $\mathbb{P}^\star$ a.s., for n large enough. This follows from the fact that, under these assumptions, denoting

$$\Delta_n(q,q^\star) = \sup_{f\in\mathscr{M}_q} \ell_n(f) - \sup_{f\in\mathscr{M}_{q^\star}} \ell_n(f),$$

$\Delta_n(q,q^\star)/n$ converges $\mathbb{P}^\star$ a.s. to minus the infimum of the Kullback distance between $f^\star$ and f, which is strictly negative if $q < q^\star$ and equal to zero if $q \geq q^\star$. In the remainder of this chapter, we will thus be interested in devising penalties guaranteeing that $\mathbb{P}^\star$ a.s., for n large enough, $\widehat{q}_n \leqslant q^\star$. Let

$$\Gamma_n(q) = \sup_{f\in\mathscr{M}_q} \ell_n(f) - \ell_n(f^\star),$$

so that $\Delta_n(q,q^\star) \leqslant \Gamma_n(q)$. The goal, when devising penalties, is to obtain an upper bound on $\Delta_n(q,q^\star)$ uniformly in $q > q^\star$.

We start with an asymptotic study of $\Gamma_n(q)$ (Sect. 4.3.4, Theorem 4.5).

This result holds under an assumption on the bracketing entropy's behavior over the set of normalized scores. We explain how things work for population mixture models in Sect. 4.3.5: why it is hard, and what we can show.

Then (Sect. 4.3.6, Theorem 4.7), under the same assumptions on the bracketing entropy's behavior over the set of normalized scores, we determine, for $q > q^*$, $\mathbb{P}^*$-a.s.

$$V(q, q^*) = \limsup_{n \to +\infty} \frac{\Delta_n(q, q^*)}{\log \log n},$$

from which it follows that consistency requires us to have

$$\forall q > q^*, \quad \text{pen}(n, q) \geqslant \text{pen}(n, q^*) + V(q, q^*) \log \log n.$$

When $V(q, q^*) > 0$, this proves that the minimum order of growth for the penalty is $\log \log n$ (for dependency in n).

Finally (Sect. 4.3.7, Theorem 4.8), since it suffices to have

$$\sup_{q > q^*} \frac{\Delta_n(q, q^*)}{\text{pen}(n, q) - \text{pen}(n, q^*)} < 1$$

to obtain consistency, we give a condition on $\eta(q) > 0$ guaranteeing that $\mathbb{P}^*$-a.s.

$$\limsup_{n \to +\infty} \sup_{q > q^*} \frac{\Delta_n(q, q^*)}{\eta(q) \log \log n} \leqslant C \tag{4.5}$$

so that it suffices to choose

$$\text{pen}(n, q) > \text{pen}(n, q^*) + C\eta(q) \log \log n$$

to obtain consistency. Note that the uniform control in $q > q^*$ of $\Delta_n(q, q^*)/\log \log n$ is not a consequence of the functional law of the iterated logarithm. It requires a normalization that depends on q, $\eta(q)$, and thus a specific proof which is not detailed in this book. We simply give an argument which helps us to understand why this result is plausible. The assumption under which the result holds is made on a local entropy, and we show, in Sect. 4.3.8, how we may infer the behavior of the local entropy of a set from the global entropy of the set of normalized functions, in a way that is applicable to population mixtures. Note that the result (4.5) is new, even as far as regular parametric models are concerned.

4.3.3 The Regular Parametric Case

If the heuristics seen with sub-models of dimension 1 are valid, then we can expect, under good assumptions, to be able to define for all $q \geqslant q^*$, a subset $\mathscr{D}_q^0$ of the unit ball of $L^2(f^* d\mu)$ (subset of normalized scores) such that

$$\sup_{f \in \mathcal{M}_q} \ell_n(f) - \ell_n(f^\star) = \frac{1}{2} \sup_{d \in \mathscr{D}_q^0} (\nu_n(d) \vee 0)^2 + o_{\mathbb{P}^\star}(1).$$

Then the asymptotic distribution of the likelihood ratio statistics between two values q_1 and q_2 such that $q_1 > q_2 \geqslant q^\star$ will be

$$\frac{1}{2}\Big\{ \sup_{d \in \mathscr{D}_{q_1}^0} (W(d) \vee 0)^2 - \sup_{d \in \mathscr{D}_{q_2}^0} (W(d) \vee 0)^2 \Big\}.$$

We then recognize the classical parametric result for nested identifiable regular parametric models. Indeed, assume that models are nested, regular parametric, and with respective dimensions $m_1 \geqslant m_2$, so that in both cases, the "true" parameter (of $f^\star$) is inside the parameter domain. Let then $E_1 \subset E_2$ be the sub-spaces of $L^2(f^\star d\mu)$, with respective dimensions m_1 and m_2, generated by the score functions in each of the models $\mathcal{M}_{q_1}$ and $\mathcal{M}_{q_2}$. Then $\mathscr{D}_{q_i}^0$ is exactly equal to the unit ball of E_i, for $i = 1, 2$. Let now $(e_1, \dots, e_{m_1})$ be an orthonormal basis of E_1 such that $(e_1, \dots, e_{m_2})$ is an orthonormal basis of E_2. Then $(W(d))_{d \in E_i}$ has the same distribution as $(\sum_{j=1}^{m_i} a_j W(e_j))_{(a_1,\dots,a_{m_i}) \in \mathbb{R}^{m_i}}$ and $(W(d))_{d \in \mathscr{D}_{q_i}^0}$ has the same distribution as

$$\Big(\sum_{j=1}^{m_i} a_j W(e_j) \Big)_{(a_1,\dots,a_{m_i}) \in \mathbb{R}^{m_i}, \sum_{j=1}^{m_i} a_j^2 = 1} \qquad (i = 1, 2).$$

On the other hand, since $W(-d) = -W(d)$ a.s., we may remove "$\vee 0$" from the formula, and twice the likelihood ratio statistics converges in distribution to

$$\sup_{\substack{(a_1,\dots,a_{m_1}) \in \mathbb{R}^{m_1} \\ \sum_{j=1}^{m_1} a_j^2 = 1}} \Big(\sum_{j=1}^{m_i} a_j W(e_j) \Big)^2 - \sup_{\substack{(a_1,\dots,a_{m_2}) \in \mathbb{R}^{m_2} \\ \sum_{j=1}^{m_2} a_j^2 = 1}} \Big(\sum_{j=1}^{m_2} a_j W(e_j) \Big)^2,$$

which is equal to $\sum_{j=m_2+1}^{m_1} (W(e_j))^2$ and is distributed as $\chi^2(m_1 - m_2)$.

In the regular parametric case, the problem reduces to subsets of finite-dimensional sub-spaces. The case of population mixtures cannot be analyzed in such a simple way: the set of normalized scores generally has infinite dimension. To see why, let us look at the simplest case where $q^\star = 1$ and $q = 2$. Take $f^\star = g_0$ and densities of $\mathcal{M}_2$ of the form

$$(1 - p) g_{\gamma_1} + p g_{\gamma_2}, \quad 0 \leqslant p \leqslant 1, \ \gamma_i \in \Gamma, \ i = 1, 2.$$

Taking as sub-model, with $\eta \in \mathbb{R}^d$, $\gamma \in \Gamma$ and $p \in [0, 1]$ fixed:

$$f_t = (1 - tp) g_{t\eta} + tp g_\gamma$$

we get the score function

$$\left\langle \eta, \frac{\dot{g}_0}{g_0} \right\rangle + p \frac{g_\gamma - g_0}{g_0},$$

where $\langle \cdot, \cdot \rangle$ is the usual scalar product in $\mathbb{R}^d$, and where we assume that the parametric model $(g_\gamma)_{\gamma \in \Gamma}$ is regular in 0 with score $\dot{g}_0/g_0$. The expected set of normalized scores has to contain all those normalized functions, i.e. divided by their norm in $L^2(f^\star d\mu)$, for all $\eta \in \mathbb{R}^d$, $p \in [0, 1]$, $\gamma \in \Gamma$.

4.3.4 Weak Approximation of the Likelihood Ratio and Asymptotic Distribution

In this section, we give a meaning to the heuristics introduced with sub-models of dimension 1, for a fixed model $\mathcal{M}_q$, and obtain an approximation, in probability, for the likelihood ratio statistics.

Let $q \geqslant q^\star$. For all density $f \in \mathcal{M}$, $f \neq f^\star$, let

$$d_f = \frac{\sqrt{f/f^\star} - 1}{h(f, f^\star)},$$

where $h(\cdot, \cdot)$ is the Hellinger distance, i.e. for probability densities f, g with respect to μ over $\mathcal{X}$, $h^2(f, g) = \int (\sqrt{f} - \sqrt{g})^2 d\mu$. We then have

$$E\left[d_f(X_1)\right] = -\frac{1}{2}h(f, f^\star) \quad \text{and} \quad E\left[d_f^2(X_1)\right] = 1.$$

For all $\varepsilon > 0$, let

$$\mathscr{D}_q(\varepsilon) = \left\{ d_f : f \in \mathcal{M}_q, \, h(f, f^\star) \leqslant \varepsilon \right\}, \quad \mathscr{D}_q = \bigcup_{\varepsilon > 0} \mathscr{D}_q(\varepsilon).$$

The possible normalized scores are the limits, in $L^2(f^\star d\mu)$, of elements of $\mathscr{D}_q(\varepsilon)$ when ε tends to 0. Let $\overline{\mathscr{D}_q(\varepsilon)}$ be the closure of $\mathscr{D}_q(\varepsilon)$ in $L^2(f^\star d\mu)$ and define

$$\mathscr{D}_q^0 = \bigcap_{\varepsilon > 0} \overline{\mathscr{D}_q(\varepsilon)}.$$

We consider normalized scores induced by sub-models of dimension 1: if $(f_t)_{t \in (0, \delta]}$, $\delta > 0$, is a sub-model such that $t \mapsto h(f_t, f^\star)$ is continuous and tends to 0 as t tends to 0, and that d_{f_t} tends to d in $L^2(f^\star d\mu)$, we let $\mathbb{D}_q$ be the set of d obtained this way. By construction, $\mathbb{D}_q \subset \mathscr{D}_q^0$.

The following two assumptions are fundamental:

$$(H1) \quad \int_0^1 \sqrt{\log N\left(\mathscr{D}_q, u\right)} \, du < +\infty,$$

$$(H2) \quad \overline{\mathbb{D}_q} = \overline{\mathscr{D}_q^0}.$$

Theorem 4.5 *Under assumptions (H1) and (H2), if $q \geq q^*$, then*

$$\sup_{f \in \mathscr{M}_q} \ell_n(f) - \ell_n\left(f^*\right) = \frac{1}{2} \sup_{d \in \mathscr{D}_q^0} \left(\nu_n(d) \vee 0\right)^2 + o_{\mathbb{P}^*}(1).$$

Remark 4.2 From this theorem, we may deduce the asymptotic distribution of the likelihood ratio statistics, under H_0, for the test of H_0: "the order is q^*" versus H_1: "the order is q". Under assumptions (H1) and (H2), the asymptotic distribution of

$$\sup_{f \in \mathscr{M}_q} \ell_n(f) - \sup_{f \in \mathscr{M}_q^*} \ell_n(f)$$

is then

$$\frac{1}{2} \left\{ \sup_{d \in \mathscr{D}_q^0} \left(W(d) \vee 0\right)^2 - \sup_{d \in \mathscr{D}_{q^*}^0} \left(W(d) \vee 0\right)^2 \right\}.$$

We may also evaluate the asymptotic power under contiguous alternatives. Let $(f_n)_{n \geqslant 1}$ be a sequence of elements of $\mathscr{M}_q$ such that $\sqrt{n} h\left(f_n, f^*\right) \to \frac{1}{2} c > 0$ and d_{f_n} tends to $d_0 \in \mathscr{D}_q^0$ in $L^2(f^* d\mu)$. We can show, in a similar way, that

$$\ell_n(f_n) - \ell_n\left(f^*\right) = c \nu_n(d_0) - \frac{1}{2} c^2 + o_{\mathbb{P}^*}(1),$$

which proves that $(f_n \mu)^{\otimes n}$ and $(f^* \mu)^{\otimes n}$ are mutually contiguous. By Le Cam's Third Lemma, we deduce from Theorem 4.5 that under $(f_n \mu)^{\otimes n}$, $\sup_{f \in \mathscr{M}_q} \ell_n(f) - \ell_n(f^*)$ converges in distribution to

$$\frac{1}{2} \sup_{d \in \mathscr{D}_q^0} \left[\left(W(d) + c \langle d, d_0 \rangle\right) \vee 0 \right]^2.$$

For notions of contiguity and Le Cam's Third Lemma, see [10]. To apply this theorem to population mixture models, we need to identify the set $\mathscr{D}_q^0$ and to show that assumptions (H1) and (H2) are verified. This is what is done below. In particular, we will have to evaluate bracketing entropies for classes of normalized scores. We will see why this is a hard problem, and how to solve it for population mixture models.

Proof (Proof of Theorem 4.5). Let us start with some simple inequalities that will be used several times. For all $f \in \mathcal{M}_q$,

$$\ell_n(f) - \ell_n(f^\star) = \sum_{i=1}^{n} 2 \log \left(1 + h\left(f, f^\star\right) d_f(X_i)\right) \tag{4.6}$$

$$\leqslant \sum_{i=1}^{n} 2h\left(f, f^\star\right) d_f(X_i) = 2\nu_n\left(d_f\right) h\left(f, f^\star\right) \sqrt{n} - h\left(f, f^\star\right)^2 n$$

and since $q \geqslant q^\star$, $\sup_{f \in \mathcal{M}_q} \ell_n(f) - \ell_n(f^\star) \geqslant 0$. If f is such that $\ell_n(f) - \ell_n(f^\star) \geqslant 0$, then $2\nu_n\left(d_f\right) h\left(f, f^\star\right) \sqrt{n} - h(f, f^\star)^2 n \geq 0$. Hence, either $f = f^\star$, in which case $h\left(f, f^\star\right) = 0$, or $\ell_n(f) - \ell_n(f^\star) > 0$, and then $h\left(f, f^\star\right) \leqslant (2/\sqrt{n})\nu_n\left(d_f\right)$. Hence

$$\sup_{\substack{f \in \mathcal{M}_q \\ \ell_n(f) - \ell_n(f^\star) \geqslant 0}} h\left(f, f^\star\right) \leqslant 0 \vee \frac{2}{\sqrt{n}} \sup_{d \in \mathcal{D}_q} \nu_n(d). \tag{4.7}$$

Now we use the expansion $2 \log(1 + x) = 2x - x^2 + x^2 R(x)$, where $R(x) \to 0$ as $x \to 0$. We obtain, for all $f \in \mathcal{M}_q$,

$$\ell_n(f) - \ell_n(f^\star) = 2h\left(f, f^\star\right) \sum_{i=1}^{n} d_f(X_i) - h\left(f, f^\star\right)^2 \sum_{i=1}^{n} d_f(X_i)^2$$

$$+ h\left(f, f^\star\right)^2 \sum_{i=1}^{n} \left(d_f(X_i)\right)^2 R\left(h\left(f, f^\star\right) d_f(X_i)\right)$$

$$= 2\sqrt{n} h\left(f, f^\star\right) \nu_n\left(d_f\right) - nh\left(f, f^\star\right)^2 - h\left(f, f^\star\right)^2 \sum_{i=1}^{n} d_f(X_i)^2$$

$$+ h\left(f, f^\star\right)^2 \sum_{i=1}^{n} d_f(X_i)^2 R\left(h\left(f, f^\star\right) d_f(X_i)\right).$$

Defining $Z_n = \sup_{d \in \mathcal{D}_q} \nu_n(d)$, $\widetilde{R}(u) = \sup_{|x| \leqslant u} |R(x)|$ and making use of (4.7), we have

$$\sup_{\substack{f \in \mathcal{M}_q : \ell_n \\ (f) - \ell_n(f^\star) \geqslant 0}} h\left(f, f^\star\right)^2 \sum_{i=1}^{n} d_f(X_i)^2 R\left(h\left(f, f^\star\right) d_f(X_i)\right)$$

$$\leqslant \frac{4Z_n^2}{n} \left[\sup_{d \in \mathcal{D}_q} \sum_{i=1}^{n} d(X_i)^2 \right] \widetilde{R}\left(\frac{2Z_n}{\sqrt{n}} \max_{1 \leqslant i \leqslant n} D(X_i)\right),$$

where D is an envelop function of $\mathcal{D}_q$ in $L^2(f^\star d\mu)$, i.e. such that $\forall d \in \mathcal{D}_q$, $|d| \leqslant D$ (D exists thanks to (H1)). But since $D \in L^2(f^\star d\mu)$,

$$\max_{1 \leqslant i \leqslant n} D\left(X_i\right) = o_{\mathbb{P}^*}\left(\sqrt{n}\right).$$

On the other hand, (H1) entails $Z_n = O_{\mathbb{P}^*}(1)$. Thus,

$$\tilde{R}\left(\frac{2Z_n}{\sqrt{n}} \max_{1 \leqslant i \leqslant n} D\left(X_i\right)\right) = o_{\mathbb{P}^*}(1).$$

Furthermore, by (H1), $\{d^2, d \in \mathscr{D}_q\}$ has a finite bracketing entropy in $L^1(f^\star d\mu)$ for all $\varepsilon > 0$, hence it is a Glivenko–Cantelli class, hence

$$\frac{1}{n} \sup_{d \in \mathscr{D}_q} \sum_{i=1}^{n}\left[d\left(X_i\right)^2 - 1\right] = o_{\mathbb{P}^*}(1),$$

and

$$\frac{4Z_n^2}{n}\left[\sup_{d \in \mathscr{D}_q} \sum_{i=1}^{n} d\left(X_i\right)^2\right] = O_{\mathbb{P}^*}(1).$$

If $\ell_n(f) - \ell_n(f^\star) \geq 0$, $h(f, f^\star) \leqslant 0 \vee \frac{2}{\sqrt{n}} \sup_{d \in \mathscr{D}_q} \nu_n(d) = O_{\mathbb{P}^*}\left(\frac{1}{\sqrt{n}}\right)$, hence

$$\sup_{\substack{f \in \mathscr{M}_q \\ \ell_n(f)-\ell_n(f^\star) \geq 0}} \left[nh^2\left(f, f^\star\right) \frac{1}{n} \sum_{i=1}^{n}\left[d_f(X_i)^2 - 1\right]\right] = o_{\mathbb{P}^*}(1),$$

which gives

$$\sup_{\substack{f \in \mathscr{M}_q \\ \ell_n(f)-\ell_n(f^\star)}} = \sup_{\substack{f \in \mathscr{M}_q \\ \ell_n(f)-\ell_n(f^\star) \geq 0}} \left[2\sqrt{n}h\left(f, f^\star\right) \nu_n\left(d_f\right) - 2nh\left(f, f^\star\right)^2\right] + o_{\mathbb{P}^*}(1).$$

Let

$$\mathscr{G}_n = \left\{f \in \mathscr{M}_q : h\left(f, f^\star\right) \leqslant n^{-1/4}\right\}.$$

Using again $h(f, f^\star) \leqslant 0 \vee \frac{2}{\sqrt{n}} \sup_{d \in \mathscr{D}_q} \nu_n(d)$ when $\ell_n(f) - \ell_n(f^\star) \geq 0$, and the fact that $\frac{2}{\sqrt{n}} \sup_{d \in \mathscr{D}_q} \nu_n(d) = O_{\mathbb{P}^*}\left(\frac{1}{\sqrt{n}}\right)$, we obtain

$$\sup_{f \in \mathscr{M}_q} \ell_n(f) - \ell_n\left(f^\star\right) = \sup_{f \in \mathscr{G}_n} \left[2\sqrt{n}h\left(f, f^\star\right) \nu_n\left(d_f\right) - 2nh\left(f, f^\star\right)^2\right] + o_{\mathbb{P}^*}(1).$$

Let Π be the orthogonal projection in $L^2(f^\star d\mu)$ on $\mathscr{D}_q^0$. Since $\mathscr{D}_q$ is compact, $\sup_{f \in \mathscr{G}_n} \|d_f - \Pi(d_f)\|$ tends to 0 when n tends to infinity, thus there exists a sequence $(u_n)_{n \in \mathbb{N}}$ tending to 0 such that

$$\sup_{f \in \mathscr{G}_n} \left\| d_f - \Pi(d_f) \right\| \leqslant u_n.$$

But, by (H1), the set of functions $d_f - \Pi(d_f)$, $f \in \mathscr{G}_n$, has a bracketing entropy integrable at 0, and an envelop function with norm upper-bounded by u_n, thus tending to 0, and by the maximal inequality (4.4)

$$\sup_{f \in \mathscr{G}_n} \nu_n \left(d_f - \Pi(d_f) \right) = o_{\mathbb{P}^*}(1),$$

so that

$$\sup_{f \in \mathscr{M}_q} \ell_n(f) - \ell_n(f^\star) = \sup_{f \in \mathscr{G}_n} \left[2\sqrt{n} h(f, f^\star) \nu_n \left(\Pi(d_f) \right) - 2nh(f, f^\star)^2 \right] + o_{\mathbb{P}^*}(1).$$

(4.8)

Note that for all n

$$\left\{ \Pi(d_f) : f \in \mathscr{G}_n \right\} = \mathscr{D}_q^0.$$

By direct optimization,

$$\sup_{f \in \mathscr{G}_n} \left[2\sqrt{n} h(f, f^\star) \nu_n \left(\Pi(d_f) \right) - 2nh(f, f^\star)^2 \right]$$

(4.9)

$$\leqslant \sup_{d \in \mathscr{D}_q^0} \sup_{p \geqslant 0} 2p\nu_n(d) - 2p^2 = \frac{1}{2} \sup_{d \in \mathscr{D}_q^0} \left(\nu_n(d) \vee 0 \right)^2.$$

But by (H2), for all $d \in \mathscr{D}_q^0$, there exists a path $(f_{\alpha,d})_{\alpha \in (0, \alpha_d]}$ such that $h(f_{\alpha,d}, f^\star) = \alpha$, with $\alpha_d > 0$ depending on d, hence for all $d \in \mathscr{D}_q^0$,

$$\sup_{f \in \mathscr{G}_n} \left[\sqrt{n} h(f, f^\star) \nu_n \left(\Pi(d_f) \right) - 2nh(f, f^\star)^2 \right]$$

$$\geqslant \sup_{\alpha \leqslant \alpha_d} 2\sqrt{n} \alpha \nu_n(d) - 2n\alpha^2 + o_{\mathbb{P}^*}(1) = \frac{1}{2} \left(\nu_n(d) \vee 0 \right)^2 + o_{\mathbb{P}^*}(1)$$

because the value of α which maximizes $2\sqrt{n} \alpha \nu_n(d) - 2n\alpha^2$ is $\nu_n(d)/2\sqrt{n} = O_{\mathbb{P}^*}(1/\sqrt{n})$.

Hence, for all finite subset S of $\mathscr{D}_q^0$,

$$\sup_{f \in \mathscr{G}_n} \left[\sqrt{n} h(f, f^\star) \nu_n \left(\Pi(d_f) \right) - 2nh(f, f^\star)^2 \right] \geqslant \frac{1}{2} \sup_{d \in S} \left(\nu_n(d) \vee 0 \right)^2 + o_{\mathbb{P}^*}(1).$$

Now, since $\mathscr{D}_q^0$ is relatively compact, appealing to (H2) and the maximal inequality (4.4), for all $\varepsilon > 0$, there exists a finite subset S of $\mathscr{D}_q^0$ such that

$$\sup_{d\in\mathscr{D}_q^0} \left(\nu_n\left(d\right)\vee 0\right)^2 \leqslant \sup_{d\in S}\left(\nu_n\left(d\right)\vee 0\right)^2 + \varepsilon$$

so that for all $\varepsilon > 0$,

$$\sup_{f\in\mathscr{G}_n}\left[\sqrt{n}\,h(f, f^\star)\nu_n\left(\Pi(d_f)\right) - 2nh\left(f, f^\star\right)^2\right] \geqslant \frac{1}{2}\sup_{d\in\mathscr{D}_q^0}\left(\nu_n\left(d\right)\vee 0\right)^2 - \varepsilon + o_{\mathbb{P}^\star}\left(1\right)$$

and finally

$$\sup_{f\in\mathscr{G}_n}\left[\sqrt{n}\,h(f, f^\star)\nu_n\left(\Pi(d_f)\right) - 2nh(f, f^\star)^2\right] \geqslant \frac{1}{2}\sup_{d\in\mathscr{D}_q^0}\left(\nu_n\left(d\right)\vee 0\right)^2 + o_{\mathbb{P}^\star}\left(1\right).$$

$$(4.10)$$

Combining (4.8), (4.9) and (4.10) yields the desired result.

4.3.5   Population Mixtures: Bracketing Entropy of Normalized Scores

To apply Theorem 4.5, the important thing is to be able to evaluate the bracketing entropy of classes $\mathscr{D}_q$ in such a way that the assumptions of the theorem are satisfied. What makes this evaluation a hard task is that the classes $\mathscr{D}_q$ are classes of normalized functions. We cannot resort to the usual regularity arguments to reduce to a finite-dimensional Euclidean space. Indeed, if the class of functions for which we try to evaluate the bracketing entropy is a parametric class of regular functions, i.e.

$$\mathscr{F} = \left\{g_\xi : \xi \in \varXi\right\}, \quad \varXi \subset \mathbb{R}^d$$

and $|g_\xi(x) - g_{\xi'}(x)| \leq G(x)\,\|\xi - \xi'\|$, where G is a function of $L^2(f^\star d\mu)$, then, to construct brackets covering $\mathscr{F}$, we can construct brackets covering the parameter set $\varXi$, which leads to

$$\mathcal{N}(\mathscr{F}, \delta) \leqslant C\left(\frac{\mathrm{Diam}(\varXi)}{\delta}\right)^d$$

for some constant $C > 0$. But, even if models $\mathscr{M}_q$ are regular parametric, there is no reason for classes $\mathscr{D}_q$ to be so, because of the denominator defining functions d_f, which may well tend to 0. We then have to answer the following:

• What is the structure of the parameter set corresponding to the set of scores at Hellinger distance upper-bounded by ε?
• How does this allow us to control the brackets of the class of normalized scores?

It is possible to answer these questions for population mixtures when the mixture is by translation. The sequel of this section aims at explaining this, and at giving an important result on the bracket entropy of the class of normalized scores, without detailing proofs.

We are in the setting where $\mathscr{X} = \mathbb{R}^d$,

$$\mathscr{M}_q = \left\{ \sum_{i=1}^{q} \pi_i g_{\gamma_i}, \ 0 \leqslant \pi_i, \ \gamma_i \in \Gamma, \ i = 1, \ldots, q, \ \sum_{i=1}^{q} \pi_i = 1 \right\},$$

Γ is a compact set of $\mathbb{R}^d$ and $g_\theta(x) = g_0(x - \theta)$ for all $x \in \mathbb{R}$. The crucial result (whose complete statement is not given here, nor its proof) to evaluate the bracketing entropy of the class of normalized scores is a tool that helps us to understand the underlying geometry. Among other things, it expresses the following: in $\mathbb{R}^d$, we can choose neighborhoods $A_1, \ldots, A_{q^\star}$ of $\gamma_1^\star, \ldots, \gamma_{q^\star}^\star$ in such a way that, if

$$A_0 = \Gamma \backslash (A_1 \cup \ldots \cup A_{q^\star}),$$

for

$$f = \sum_{i=1}^{q} \pi_i g_{\gamma_i},$$

the Hellinger distance $h(f, f^\star)$ is upper and lower bounded, up to constants that do not depend on q, by the pseudo-distance

$$\sum_{\theta_j \in A_0} \pi_j + \sum_{i=1}^{q^\star} \left\{ \left| \sum_{\gamma_j \in A_i} \pi_j - \pi_i^\star \right| + \left\| \sum_{\theta_j \in A_i} \pi_j (\gamma_j - \gamma_i^\star) \right\| + \frac{1}{2} \sum_{\gamma_j \in A_i} \pi_j \| \gamma_j - \gamma_i^\star \|^2 \right\}.$$

We may indeed see via an example that the parameters' geometry corresponding to scores inside a Hellinger ball is far from the geometry of a Euclidean ball.

Let g_0 be the density of a centered Gaussian random variable with variance $\frac{1}{4}$ over $\mathbb{R}$, $f^\star = g_{0.5}$. We have $q^\star = 1$ and $\mathscr{M}_2 = \{ pg_{\gamma_1} + (1 - p)g_2 : p, \gamma_1, \gamma_2 \in [0, 1] \}$ (Fig. 4.1).

We can then show that, nevertheless, the bracketing entropy of the set of normalized scores is analogous to that of a Euclidean ball, i.e. polynomial with an exponent proportional to the dimension, i.e. proportional to the order, with constants which do not depend on the dimension, as is the case in Euclidean geometry.

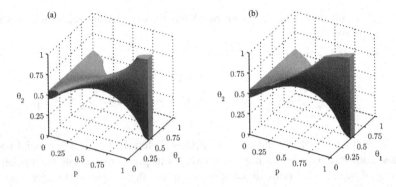

Fig. 4.1 (a) represents the set of parameters corresponding to $\{f \in \mathcal{M}_2 : h(f, f^\star) \leq 0.05\}$, and
(b) represents the set of parameters (p, θ_1, θ_2) such that $|p(\theta_1 - 0.5) + (1 - p)(\theta_2 - 0.5)| + \frac{1}{2}p(\theta_1 - 0.5)^2 + \frac{1}{2}(1 - p)(\theta_2 - 0.5)^2 \leq 0.05$

Assume that g_0 is three times continuously differentiable. Let

$$H_0(x) = \sup_{\gamma \in \Gamma} \frac{g_\gamma(x)}{f^\star(x)},$$

$$H_1(x) = \sup_{\gamma \in \Gamma} \max_{i=1,\dots,d} \frac{|\partial g_\gamma(x)/\partial \gamma^i|}{f^\star(x)},$$

$$H_2(x) = \sup_{\gamma \in \Gamma} \max_{i,j=1,\dots,d} \frac{|\partial^2 g_\gamma(x)/\partial \gamma^i \partial \gamma^j|}{f^\star(x)},$$

$$H_3(x) = \sup_{\gamma \in \Gamma} \max_{i,j,k=1,\dots,d} \frac{|\partial^3 g_\gamma(x)/\partial \gamma^i \partial \gamma^j \partial \gamma^k|}{f^\star(x)}.$$

Theorem 4.6 *Assume that:*

- $g_0 \in C^3$ *and* $g_0(x)$, $(\partial g_0/\partial \theta^i)(x)$ *tend to 0 as* $\|x\| \to \infty$;
- $H_k \in L^4(f^\star \mathrm{d}\mu)$ *for* $k = 0, 1, 2$ *and* $H_3 \in L^2(f^\star \mathrm{d}\mu)$.

Then there exist constants $C^\star$ *and* $\delta^\star$, *depending on* d, $q^\star$ *and* $f^\star$ *but not on* Γ, q *nor* δ, *such that*

$$\mathcal{N}(\mathcal{D}_q, \delta) \leq \left(\frac{C^\star(T \vee 1)^{1/6}(\|H_0\|_4^4 \vee \|H_1\|_4^4 \vee \|H_2\|_4^4 \vee \|H_3\|_2^2)}{\delta} \right)^{18(d+1)q}$$

for all $q \geq q^\star$, $\delta \leq \delta^\star$, *where* $2T$ *is the diameter of* Γ.

To prove this theorem, we consider separately normalized scores that are close to $f^\star$ and those that are far: for $\alpha > 0$, if

$$\mathcal{D}_{q,\alpha} = \left\{ d_f : f \in \mathcal{M}_q,\ f \neq f^\star,\ h(f, f^\star) \leq \alpha \right\}$$

then

$$\mathcal{N}(\mathcal{D}_q, \delta) \le \mathcal{N}(\mathcal{D}_{q,\alpha}, \delta) + \mathcal{N}(\mathcal{D}_q \setminus \mathcal{D}_{q,\alpha}, \delta).$$

Thanks to the regularity, computation of $\mathcal{N}(\mathcal{D}_q \setminus \mathcal{D}_{q,\alpha}, \delta)$ can be done by reducing to Euclidean computations in the parameter space, but to compute $\mathcal{N}(\mathcal{D}_{q,\alpha}, \delta)$, we need to understand the behavior of d_f when $h(f, f^\star)$ is small. This is where the result on the geometry of Hellinger balls steps in. Lastly, we need to carefully choose α as a function of δ. See [12].

Using the same tools, $\mathcal{D}_q^0$ can be identified. When $d = 1$, for $q > q^\star$, $\mathcal{D}_q^0$ is the set of functions of the form

$$\frac{\displaystyle\sum_{i=1}^{q^\star} \left\{ \eta_i \frac{g_{\gamma_i^\star}}{f^\star} + \beta_i \frac{D_1 g_{\gamma_i^\star}}{f^\star} \right\} + \sum_{i=1}^{r} \rho_i \frac{D_2 g_{\gamma_i^\star}}{f^\star} + \sum_{j=1}^{s} \tau_j \frac{g_{\gamma_j}}{f^\star}}{\left\| \displaystyle\sum_{i=1}^{q^\star} \left\{ \eta_i \frac{g_{\gamma_i^\star}}{f^\star} + \beta_i \frac{D_1 g_{\gamma_i^\star}}{f^\star} \right\} + \sum_{i=1}^{r} \rho_i \frac{D_2 g_{\theta_i^\star}}{f^\star} + \sum_{j=1}^{s} \tau_j \frac{g_{\gamma_j}}{f^\star} \right\|}$$

where r and s are integers such that

$$q^\star + r + s \le q,$$

and where η_i, β_i $(i = 1, \ldots, q^\star)$ are real numbers, ρ_i $(i = 1, \ldots, r)$, γ_i $(i = 1, \ldots, s)$, are non-negative real numbers, such that

$$\sum_{i=1}^{q^\star} \eta_i + \sum_{j=1}^{s} \gamma_j = 0.$$

In particular, for $q = q^\star$, we obtain

$$\mathcal{D}_{q^\star}^0 = \left\{ \frac{L}{\|L\|_2} : L = \sum_{i=1}^{q^\star} \left\{ \eta_i \frac{f_{\theta_i^\star}}{f^\star} + \beta_i \frac{D_1 f_{\theta_i^\star}}{f^\star} \right\}, \ \eta, \beta \in \mathbb{R}^{q^\star}, \ \sum_{i=1}^{q^\star} \eta_i = 0 \right\}.$$

We then notice that those functions can be obtained with a continuous sub-model. To see this, it suffices to consider sub-models of the form, for $t \ge 0$ tending to 0:

$$\sum_{i=1}^{q^\star - r} \left(\pi_i^\star + t^2 \eta_i \right) \frac{g_{\theta_i^\star}}{f^\star} + \sum_{i=1}^{r} \left(\pi_i^\star + t^2 \eta_i \right) \left[\frac{g_{\theta_i^\star + t(\rho_i + t\beta_i)}}{f^\star} + \frac{g_{\theta_i^\star - t\rho_i}}{f^\star} \right] + \sum_{j=1}^{s} t^2 \gamma_j \frac{g_{\theta_j}}{f^\star}.$$

4.3.6   Approximation, Functional Law of the Iterated Logarithm and Minimum Penalty

One may strengthen the previous approximation of the likelihood ratio, so as to obtain an almost sure limiting result, and be able to minimize the order of growth of the penalty yielding an almost surely consistent estimator for the order by penalized maximum likelihood.

Theorem 4.7 *Assume that (H1) and (H2) hold. Let $q > q^\star$ and D be an envelop function of $\mathscr{D}_q$, i.e. such that $|d| \leqslant D$ for all $d \in \mathscr{D}_q$. Assume that, for some $\alpha > 0$, we have $E D^{2+\alpha}(X_1) < +\infty$. Then $\mathbb{P}^\star$ a.s.:*

$$\limsup_{n \to +\infty} \frac{1}{\log \log n} \Big[\sup_{f \in \mathscr{M}_q} \ell_n(f) - \sup_{f \in \mathscr{M}_{q^\star}} \ell_n(f) \Big]$$
$$= \sup_{g \in L_0^2(f^\star d\mu)} \Big[\sup_{d \in \mathscr{D}_q^0} \langle d, g \rangle^2 - \sup_{d \in \mathscr{D}_{q^\star}^0} \langle d, g \rangle^2 \Big].$$

The proof of this result can be found in [13].

Thanks to this theorem, we can deduce that $\log \log n$ is the minimum order of growth for the penalty yielding a consistent estimator for the order. Let us define

$$\widehat{q}_n = \operatorname{argmax}_{q \in \mathbb{N}} \Big\{ \sup_{f \in \mathscr{M}_q} \ell_n(f) - \operatorname{pen}(n, q) \Big\}$$

with a penalty of the form

$$\operatorname{pen}(n, q) = C\eta(q) \log \log n, \tag{4.11}$$

where C is a positive constant and $\eta(\cdot)$ is a strictly increasing function over $\mathbb{N}$. Then, $\mathbb{P}^\star$ a.s.:

$$\limsup_{n \to +\infty} \frac{\sup_{f \in \mathscr{M}_q} \ell_n(f) - \sup_{f \in \mathscr{M}_{q^\star}} \ell_n(f)}{\operatorname{pen}(n, q) - \operatorname{pen}(n, q^\star)} = \frac{V(q, q^\star)}{C(\eta(q) - \eta(q^\star))}$$

with

$$V(q, q^\star) = \sup_{g \in L_0^2(f^\star d\mu)} \Big[\sup_{d \in \mathscr{D}_q^0} (\langle d, g \rangle)^2 - \sup_{d \in \mathscr{D}_{q^\star}^0} (\langle d, g \rangle)^2 \Big].$$

Hence, if $V(q, q^\star) > 0$, as soon as $C < V(q, q^\star)/(\eta(q) - \eta(q^\star))$, $\mathbb{P}^\star$ a.s.

$$\limsup_{n \to +\infty} \frac{\sup_{f \in \mathscr{M}_q} \ell_n(f) - \sup_{f \in \mathscr{M}_{q^\star}} \ell_n(f)}{\operatorname{pen}(n, q) - \operatorname{pen}(n, q^\star)} > 1,$$

thus $\mathbb{P}^\star$ a.s., for an infinity of n,

$$\sup_{f \in \mathcal{M}_q} \ell_n (f) - \operatorname{pen}(n, q) > \sup_{f \in \mathcal{M}_{q^*}} \ell_n (f) - \operatorname{pen}(n, q^*)$$

and thus $\mathbb{P}^*$ a.s., $\widehat{q}_n \neq q^*$ for an infinity of n. Finally, we have $V(q, q^*) > 0$, as soon as

$$\mathscr{D}_q^0 \backslash \overline{\mathscr{D}_{q^*}^0} \neq \emptyset.$$

Indeed, let $g \in \mathscr{D}_q^0 \backslash \overline{\mathscr{D}_{q^*}^0}$. Then, since $g \in L_0^2 (f^* \mathrm{d}\mu)$ and $g \notin \overline{\mathscr{D}_{q^*}^0}$,

$$V(q, q^*) \geqslant 1 - \sup_{d \in \mathscr{D}_{q^*}^0} \left(\langle d, g \rangle \right)^2 > 0.$$

This entails that, to obtain an almost surely consistent estimator for the order, even when a bound on the order is known *a priori*, the penalty has to scale at least as (4.11) with C large enough (greater than $V(q, q^*)/(\eta(q) - \eta(q^*))$).

4.3.7 Uniform Law of the Iterated Logarithm and Sufficient Penalty

In order to calibrate a penalty that suffices to consistently estimate the order without any *prior* bound, we have to control the fluctuations of the likelihood ratio statistics simultaneously for all models in the collection. This could be done with a uniform functional law of the iterated logarithm in all orders. However

$$\mathscr{D} = \bigcup_{q > q^*} \mathscr{D}_q$$

does not generally satisfy the conditions for the functional law of the iterated logarithm.

Nevertheless, we can establish a law of the iterated logarithm which holds uniformly for all orders q, by introducing a factor depending on the order q. To do so, we rework the proof technique for laws of the iterated logarithm thanks to deviation inequalities. We use the local bracketing entropy of Hellinger balls of non-normalized score functions. Define, for all $\varepsilon > 0$

$$\mathscr{H}_q(\varepsilon) = \left\{ \sqrt{f / f^*} : \ f \in \mathcal{M}_q, \ h(f, f^*) \leqslant \varepsilon \right\}.$$

Theorem 4.8 *Assume that there exists a $K > 0$ such that, for all $q \geqslant q^*$ and $\delta \leqslant \varepsilon$*

$$N(\mathscr{H}_q(\varepsilon), \delta) \leqslant \left(\frac{K\varepsilon}{\delta} \right)^{\eta(q)},$$

where $\eta(q) \geqslant q$ is an increasing function. Then there exists a universal constant $C > 0$ such that $\mathbb{P}^*$-a.s.

$$\limsup_{n\to\infty} \frac{1}{\log\log n} \sup_{q\geqslant q^*} \frac{1}{\eta(q)} \left\{ \sup_{f\in\mathcal{M}_q} \ell_n(f) - \sup_{f\in\mathcal{M}_{q^*}} \ell_n(f) \right\} \leqslant C.$$

From this theorem, $\log\log n$ is a sufficient order of growth for the penalty to obtain a consistent estimator of the order, without any *prior* bound. Indeed, choose a penalty of the form

$$\mathrm{pen}\,(n,q) = \eta(q)\,v(n)\log\log n.$$

Then,

$$\sup_{q>q^*} \frac{\sup_{f\in\mathcal{M}_q} \ell_n(f) - \sup_{f\in\mathcal{M}_{q^*}} \ell_n(f)}{\mathrm{pen}\,(n,q) - \mathrm{pen}\,(n,q^*)}$$

$$= \sup_{q>q^*} \frac{1}{\eta(q)} \frac{\eta(q)}{\eta(q) - \eta(q^*)} \frac{\sup_{f\in\mathcal{M}_q} \ell_n(f) - \sup_{f\in\mathcal{M}_{q^*}} \ell_n(f)}{v(n)\log\log n}$$

$$\leqslant \frac{\eta(q^*+1)}{\eta(q^*+1) - \eta(q^*)} \frac{1}{v(n)} \sup_{q>q^*} \frac{1}{\eta(q)} \frac{\sup_{f\in\mathcal{M}_q} \ell_n(f) - \sup_{f\in\mathcal{M}_{q^*}} \ell_n(f)}{\log\log n}.$$

Hence, choosing $v(n)$ such that for all n,

$$v(n) > \frac{C\eta(q^*+1)}{\eta(q^*+1) - \eta(q^*)},$$

$\mathbb{P}^*$ a.s., for n large enough, for all integers $q > q^*$,

$$\sup_{f\in\mathcal{M}_q} \ell_n(f) - \mathrm{pen}\,(n,q) < \sup_{f\in\mathcal{M}_{q^*}} \ell_n(f) - \mathrm{pen}(n,q^*)$$

and thus $\mathbb{P}^*$ a.s., for n large enough, $\widehat{q}_n \leqslant q^*$. Since q^* is unknown, one way to achieve this is to choose $v(n)$ tending to $+\infty$ with n (arbitrarily slowly).

The complete proof of Theorem 4.8 is not given here. It can be found in [13]. We will just give some ideas that help us to understand why this result is plausible.

We want to find a way to control large values of the likelihood ratio, i.e. to control the tail of this random variable. Theorem 4.5 tells us that asymptotically, for all $q > q^*$, the likelihood ratio

$$\sup_{f\in\mathcal{M}_q} \ell_n(f) - \sup_{f\in\mathcal{M}_{q^*}} \ell_n(f)$$

is lower-bounded in distribution by the square of a Gaussian random variable, whose tails decrease exponentially fast. Hence, in the best case, we may expect that for some constants $C_1, C_2 > 0$ and A large enough:

$$\mathbb{P}^{\star}\Big(\sup_{f \in \mathcal{M}_q} \ell_n(f) - \sup_{f \in \mathcal{M}_{q^{\star}}} \ell_n(f) \geqslant A \Big) \leqslant C_1 e^{-C_2 A}. \tag{4.12}$$

In order to use the Borel–Cantelli Lemma and infer almost sure results, we need a summable series, which is obtained by choosing $A = C \log n$ in this inequality with C such that $CC_2 > 1$. This entails that, $\mathbb{P}^{\star}$ a.s., for n large enough, $\sup_{f \in \mathcal{M}_q} \ell_n(f) - \sup_{f \in \mathcal{M}_{q^{\star}}} \ell_n(f) \leqslant C \log n$ and this is not enough to get a law of the iterated logarithm.

Considering Theorem 4.5, one cannot hope to obtain a faster decay, for tails of the likelihood ratio, than a decay of the form (4.12). Nevertheless, it is possible to obtain a uniform law of the iterated logarithm thanks to an exponential deviation inequality similar to (4.12). Let us explain how.

For all n and all $q > q^{\star}$ let $Z_n(q)$ be a random variable such that there exists a function $d(\cdot)$ over $\mathbb{N}$ and a real number $\gamma > 0$ such that for all integers q,

$$d(q+1) - d(q) \geqslant \gamma \tag{4.13}$$

and for all integers n

$$\forall A \geqslant d(q), \quad \mathbb{P}^{\star}\Big(\max_{k \leqslant n} Z_k(q) \geqslant A \Big) \leqslant C_1 e^{-C_2 A}. \tag{4.14}$$

Note that if $Z_n(q)$ is the square of an empirical process, inequality (4.14) can be obtained by inequalities for $\mathbb{P}^{\star}(Z_k(q) \geqslant A)$ for all $k \leqslant n$ thanks to an analogue of Etemadi's Inequality, see Proposition 4.9 below. On the other hand, (4.13) tells us that $d(q)$ grows at least as γq. If this quantity is seen as an approximation for the expectation of $Z_n(q)$, one may hope that $d(q)$ is of order q if this random variable is approximately distributed as a $\chi^2(q)$.

Then, if $C \geqslant 1$, for all integers n:

$$\mathbb{P}^{\star}\Big(\max_{2^n \leqslant k \leqslant 2^{n+1}} \frac{Z_k(q)}{d(q) \log \log k} \geqslant C \Big) \leqslant \mathbb{P}^{\star}\Big(\max_{2^n \leqslant k \leqslant 2^{n+1}} \frac{Z_k(q)}{\log \log 2^n} \geqslant Cd(q) \Big)$$

$$\leqslant \mathbb{P}^{\star}\Big(\max_{k \leqslant 2^{n+1}} Z_k(q) \geqslant Cd(q) \log \log 2^n \Big)$$

$$\leqslant C_1 e^{-C_2 Cd(q) \log \log 2^n}.$$

Summing over $q > q^{\star}$ and using (4.13), we get, for some constant $C_3 > 0$,

$$\mathbb{P}^{\star}\Big(\sup_{q > q^{\star}} \max_{2^n \leqslant k \leqslant 2^{n+1}} \frac{Z_k(q)}{d(q) \log \log k} \geqslant C \Big) \tag{4.15}$$

$$\leqslant C_3 e^{-C_2 C\gamma \log \log 2^n} = C_3 e^{-C_2 C\gamma \log \log 2} n^{-C_2 C\gamma},$$

which is summable in n as soon as $C_2 C\gamma > 1$. The Borel–Cantelli Lemma then implies that, $\mathbb{P}^{\star}$ a.s.,

$$\limsup_{n \to +\infty} \frac{1}{\log\log n} \sup_{q > q^*} \frac{Z_n(q)}{d(q)} \leqslant C. \tag{4.16}$$

Proposition 4.9 *Let $\mathscr{Q}$ be a family of measurable functions $h : \mathscr{X} \to \mathbb{R}$. Then for all $\alpha > 0$ and $m, n \in \mathbb{N}$ such that $m \leq n$,*

$$\mathbb{P}^* \left[\max_{k=m,\ldots,n} \sup_{h \in \mathscr{Q}} |S_k(h)| \geq 3\alpha \right] \leq 3 \max_{k=m,\ldots,n} \mathbb{P}^* \left[\sup_{h \in \mathscr{Q}} |S_k(h)| \geq \alpha \right],$$

with $S_n(h) = \sum_{i=1}^{n} h(X_i)$.

Proof (Proof of Proposition 4.9). The proof follows that of the classical Etemadi's Inequality. Define the stopping time

$$\tau = \inf \left\{ k \geq m : \sup_{h \in \mathscr{Q}} |S_k(h)| \geq 3\alpha \right\}.$$

We have

$$\mathbb{P}^* \left[\max_{k=m,\ldots,n} \sup_{h \in \mathscr{Q}} |S_k(h)| \geq 3\alpha \right] = \mathbb{P}^*[\tau \leq n]$$

$$\leq \mathbb{P}^* \left[\sup_{h \in \mathscr{Q}} |S_n(h)| \geq \alpha \right] + \sum_{k=m}^{n} \mathbb{P}^* \left[\tau = k \text{ and } \sup_{h \in \mathscr{Q}} |S_n(h)| < \alpha \right].$$

But on the event $\{\tau = k \text{ and } \sup_{h \in \mathscr{Q}} |S_n(h)| < \alpha\}$, we have

$$2\alpha \leq \sup_{h \in \mathscr{Q}} |S_k(h)| - \sup_{h \in \mathscr{Q}} |S_n(h)| \leq \sup_{h \in \mathscr{Q}} |S_k(h) - S_n(h)|,$$

so that

$$\mathbb{P}^* \left[\max_{k=m,\ldots,n} \sup_{h \in \mathscr{Q}} |S_k(h)| \geq 3\alpha \right] \leq \mathbb{P}^* \left[\sup_{h \in \mathscr{Q}} |S_n(h)| \geq \alpha \right]$$

$$+ \sum_{k=m}^{n} \mathbb{P}^* \left[\tau = k \text{ and } \sup_{h \in \mathscr{Q}} |S_n(h) - S_k(h)| \geq 2\alpha \right].$$

Since $\sup_{h \in \mathscr{Q}} |S_n(h) - S_k(h)|$ and $\{\tau = k\}$ are independent, we obtain

$$\mathbb{P}^* \left[\max_{k=m,\ldots,n} \sup_{h \in \mathscr{Q}} |S_k(h)| \geq 3\alpha \right] \leq \mathbb{P}^* \left[\sup_{h \in \mathscr{Q}} |S_n(h)| \geq \alpha \right]$$

$$+ \max_{k=m,\ldots,n} \mathbb{P}^* \left[\sup_{h \in \mathscr{Q}} |S_n(f) - S_k(h)| \geq 2\alpha \right],$$

and the proposition easily follows.

4.3.8 Population Mixtures: Local Bracketing Entropy of Scores

Theorem 4.8 requires a control on the local bracketing entropy of Hellinger balls of non-normalized score functions. We observed that the geometry of Hellinger balls does not look like the Euclidean geometry of the parameter space.

Using the classical Euclidean reduction and focusing on parametric classes of regular functions g_ξ such that $|g_\xi(x) - g_{\xi'}(x)| \leq G(x) \|\xi - \xi'\|$ where G is a function of $L^2(f^* d\mu)$, letting, for ξ^* fixed and $\varepsilon > 0$,

$$\mathscr{F}(\varepsilon) = \left\{ g_\xi - g_{\xi^*} : \xi \in \Xi, \ \|g_\xi - g_{\xi^*}\|_2 \leqslant \varepsilon \right\},$$

and further assuming that $\|g_\xi - g_{\xi^*}\|_2 \geqslant c\|\xi - \xi^*\|$, then the local bracketing entropy is polynomial in ε/δ: Euclidean computations show that there exists a constant $C > 0$ such that, if K is the diameter of Ξ,

$$\mathscr{N}\left(\mathscr{F}(\varepsilon), \delta\right) \leqslant C\left(K\frac{\varepsilon}{\delta}\right)^d.$$

Similarly, we can compute the bracketing entropy of the set of normalized scores. Computation of the local bracketing entropy of non-normalized scores then requires us to understand the local geometry in a Hellinger neighborhood of f^*.

When the behavior of the bracketing entropy of the set of normalized scores is polynomial, one may infer the behavior of the local bracketing entropy of non-normalized scores. This is what is expressed in the following general theorem.

Let $\mathscr{M}$ be a set of densities over $\mathscr{X}$ with respect to μ and

$$\mathscr{D} = \left\{ d_f : f \in \mathscr{M}, \ f \neq f^* \right\}.$$

For all $\delta > 0$, define the Hellinger ball

$$\mathscr{H}(\delta) = \left\{ \sqrt{f/f^*} : h(f, f^*) \leq \delta \right\}.$$

We then have the following.

Theorem 4.10 *Assume that there exist* $q, C_0 \geq 1$ *and* $\varepsilon_0 > 0$ *such that*

$$\mathscr{N}(\mathscr{D}, \varepsilon) \leq \left(\frac{C_0}{\varepsilon}\right)^q \quad \text{for every } \varepsilon \leq \varepsilon_0.$$

Let $R \geq \sup_f |d_f|$ *be an envelop function such that* $\|R\|_2 < \infty$. *Then*

$$\mathscr{N}(\mathscr{H}(\delta), \rho) \leq \left(\frac{C_1\delta}{\rho}\right)^{q+1}.$$

for all $\delta, \rho > 0$ such that $\rho/\delta < 4 \wedge 2\|R\|_2$, with $C_1 = 8C_0(1 \vee \|R\|_2/4\varepsilon_0)$.

Proof (Proof of Theorem 4.10). The assumptions imply that

$$\mathcal{N}(\mathscr{D}, \varepsilon) \leq \left(\frac{C_0}{\varepsilon \wedge \varepsilon_0}\right)^q \quad \text{for all } \varepsilon > 0.$$

If $\varepsilon < \frac{1}{4}\|R\|_2$, then $\varepsilon/\varepsilon \wedge \varepsilon_0 \leq 1 \vee \|R\|_2/4\varepsilon_0$. Letting $C = C_0(1 \vee \|R\|_2/4\varepsilon_0)$, we get

$$\mathcal{N}(\mathscr{D}, \varepsilon) \leq \left(\frac{C}{\varepsilon}\right)^q \quad \text{for every } \varepsilon < \frac{1}{4}\|R\|_2. \tag{4.17}$$

We now prove that for all $\delta, \rho > 0$ such that $\rho/\delta < 4 \wedge 2\|R\|_2$, we have

$$\mathcal{N}(\mathscr{H}(\delta), \rho) \leq \left(\frac{8C\delta}{\rho}\right)^{q+1}.$$

Let $\varepsilon, \delta > 0$ and $N = \mathcal{N}(\mathscr{D}, \varepsilon)$. Then there exist $\ell_1, u_1, \ldots, \ell_N, u_N$ such that $\|u_i - \ell_i\|_2 \leq \varepsilon$ for all i, and for all f, there exists an i such that $\ell_i \leq d_f \leq u_i$. When f is such that $r^{-n}\delta \leq h(f, f^\star) \leq r^{-n+1}\delta$ (with $r > 1$), there exists an i such that

$$(r^{-n}\ell_i \wedge r^{-n+1}\ell_i)\delta + 1 \leq \sqrt{f/f^\star} \leq (r^{-n}u_i \vee r^{-n+1}u_i)\delta + 1.$$

Note that

$$\|u_i\, r^{-n}\delta - \ell_i\, r^{-n}\delta\|_2 \leq r^{-n}\delta\varepsilon,$$
$$\|u_i\, r^{-n+1}\delta - \ell_i\, r^{-n+1}\delta\|_2 \leq r^{-n+1}\delta\varepsilon,$$
$$\|u_i\, r^{-n+1}\delta - \ell_i\, r^{-n}\delta\|_2 \leq (r-1)r^{-n}\delta + r^{-n+1}\delta\varepsilon,$$
$$\|u_i\, r^{-n}\delta - \ell_i\, r^{-n+1}\delta\|_2 \leq (r-1)r^{-n}\delta + r^{-n+1}\delta\varepsilon,$$

where the last two inequalities come from $\ell_i \leq d_f \leq u_i$, $\|d_f\|_2 = 1$, and

$$(u_i - \ell_i)\, r^{-n}\delta \leq u_i\, r^{-n+1}\delta - \ell_i\, r^{-n}\delta - d_f\, (r-1)r^{-n}\delta \leq (u_i - \ell_i)\, r^{-n+1}\delta,$$
$$(u_i - \ell_i)\, r^{-n}\delta \leq u_i\, r^{-n}\delta - \ell_i\, r^{-n+1}\delta + d_f\, (r-1)r^{-n}\delta \leq (u_i - \ell_i)\, r^{-n+1}\delta.$$

Since $|a \vee b - c \wedge d| \leq |a - c| + |a - d| + |b - c| + |b - d|$, we have

$$\left\|(r^{-n}u_i \vee r^{-n+1}u_i)\delta - (r^{-n}\ell_i \wedge r^{-n+1}\ell_i)\delta\right\|_2 \leq 2(r-1)r^{-n}\delta + 4r^{-n+1}\delta\varepsilon.$$

Altogether, we have shown that for all $\varepsilon, \delta > 0, r > 1, n \in \mathbb{N}$,

$$\mathcal{N}\left(\{\sqrt{f/f^\star} : r^{-n}\delta \leq h(f, f^\star) \leq r^{-n+1}\delta\}, 2(r-1)r^{-n}\delta + 4r^{-n+1}\delta\varepsilon\right) \leq \mathcal{N}(\mathscr{D}, \varepsilon).$$

In particular, for all $\delta > 0, r > 1, n \in \mathbb{N}, \rho > 2(r-1)r^{-n}\delta$,

$$\mathcal{N}\left(\left\{\sqrt{f/f^\star} : r^{-n}\delta \le h(f, f^\star) \le r^{-n+1}\delta\right\}, \rho\right) \le \mathcal{N}\left(\mathcal{D}, \tfrac{1}{4}r^{n-1}\rho/\delta - \tfrac{1}{2}(1 - 1/r)\right).$$

As soon as $h(f, f^\star) \le r^{-n}\delta$, we have

$$1 - r^{-n}\delta R \le \sqrt{f/f^\star} \le 1 + r^{-n}\delta R.$$

It follows that

$$\mathcal{N}\left(\left\{\sqrt{f/f^\star} : h(f, f^\star) \le r^{-\lceil H \rceil}\delta\right\}, 2r^{-H}\delta\|R\|_2\right) = 1$$

for all $\delta > 0$, $r > 1$, $H > 0$. Consequently,

$$\mathcal{N}\left(\left\{\sqrt{f/f^\star} : h(f, f^\star) \le \delta\right\}, 2r^{-H}\delta\|R\|_2\right)$$

$$\le 1 + \sum_{n=1}^{\lceil H \rceil} \mathcal{N}\left(\left\{\sqrt{f/f^\star} : r^{-n}\delta \le h(f, f^\star) \le r^{-n+1}\delta\right\}, 2r^{-H}\delta\|R\|_2\right)$$

$$\le 1 + \sum_{n=1}^{\lceil H \rceil} \mathcal{N}\left(\mathcal{D}, \frac{1}{2}\{r^{n-H-1}\|R\|_2 - (1 - 1/r)\}\right)$$

as soon as $\delta > 0$, $r > 1$, $H > 0$ are such that $\|R\|_2 > (1 - 1/r)r^H$. In particular,

$$\mathcal{N}\left(\left\{\sqrt{f/f^\star} : h(f, f^\star) \le \delta\right\}, 2r^{-H}\delta\|R\|_2\right) \le 1 + \sum_{n=1}^{\lceil H \rceil} \mathcal{N}\left(\mathcal{D}, \tfrac{1}{4}r^{n-H-1}\|R\|_2\right)$$

as soon as $\delta > 0$, $r > 1$, $H > 0$ are such that $\|R\|_2 \ge 2(1 - 1/r)r^H$, using the entropy decay with respect to the bracket size. Thanks to (4.17),

$$\mathcal{N}\left(\left\{\sqrt{f/f^\star} : h(f, f^\star) \le \delta\right\}, 2r^{-H}\delta\|R\|_2\right) \le 1 + \sum_{n=1}^{\lceil H \rceil} r^{-(n-1)q}\left(\frac{8C}{2r^{-H}\|R\|_2}\right)^q$$

as soon as $\delta > 0$, $r > 1$, $H > 0$ satisfy $\|R\|_2 \ge 2(1 - 1/r)r^H$. But for $r > 1$ and $q, C \ge 1$

$$\sum_{n=1}^{\lceil H \rceil} r^{-(n-1)q} \le \frac{1}{1 - 1/r^q} \le \frac{1}{1 - 1/r} \le \frac{\|R\|_2}{2(1 - 1/r)r^H} \frac{4C}{2r^{-H}\|R\|_2}.$$

We thus obtain

$$\mathcal{N}\left(\left\{\sqrt{f/f^\star} : h(f, f^\star) \le \delta\right\}, 2r^{-H}\delta\|R\|_2\right) \le \frac{\|R\|_2}{2(1 - 1/r)r^H}\left(\frac{8C}{2r^{-H}\|R\|_2}\right)^{q+1}$$

as soon as $\delta > 0, r > 1, H > 0$ are such that $\|R\|_2 \geq 2(1 - 1/r)r^H$.

Let us now fix $\delta, \rho > 0$ such that $\rho/\delta < 4 \wedge 2\|R\|_2$ and choose

$$r = \frac{4}{4 - \rho/\delta}, \quad H = \frac{\log(2\|R\|_2\delta/\rho)}{\log r}.$$

We have $r > 1$ and $H > 0$. Moreover, this choice of r and H implies $\|R\|_2 = 2(1 - 1/r)r^H$ and $\rho = 2r^{-H}\delta\|R\|_2$. We have shown that

$$\mathscr{N}\left(\{\sqrt{f/f^\star} : h(f, f^\star) \leq \delta\}, \rho\right) \leq \left(\frac{8C\delta}{\rho}\right)^{q+1}$$

for all $\delta, \rho > 0$ such that $\rho/\delta < 4 \wedge 2\|R\|_2$.

4.4 Notes

In the case of Markov chains with finite state space, Ciszar and Shields [14] showed that the BIC penalty was sufficient to identify the Markovian order, using the explicit form of the maximum likelihood. Still in the case of Markov chains with finite state space, Ramon van Handel [4] showed that $\log \log n$ was a sufficient penalty to obtain an almost surely consistent order estimator (without any *prior* bound), using deviation inequalities for the empirical process, instead of the explicit form of the maximum likelihood.

The study of hidden Markov chains with values in a finite alphabet comes from [7].

The study of hidden Markov chains with Gaussian emission comes from [8], where hidden Markov chains with Poissonian emission are also investigated.

One may also be interested in hidden Markov chains for which the underlying Markov chain is not of order 1, but is a context tree source. The order then corresponds to its context tree. Models are not nested anymore. But it is still possible to use universal coding techniques to estimate the underlying context tree, see [15].

The study of likelihoods for sequences of independent variables, with applications to population mixtures, comes from [16] and [17], as far as the asymptotic distribution of the likelihood ratio is concerned, and from [13], as far as almost sure consistent order estimation and minimum sufficient penalty evaluation are concerned. The result on bracketing entropies for population mixtures and the method allowing us to infer a local entropy from the entropy of a normalized class can be found in the article [12]. The precise results stated in [13] and in [12] are more general than those stated in this book. In particular, the uniform law of the iterated logarithm is written for models which can also be increasing in n, the number of observations. Among other things, this applies to parametric classes with unbounded parameters, and growth with respect to the parameter appears in the normalization of the uniform law of the iterated logarithm.

Evaluation of the error probability in order estimation relies on deviations of the likelihood ratio: the under-estimation probability relies on large deviation properties of the likelihood ratio statistics, whereas the over-estimation probability relies on moderate deviation properties of this statistics. In the case of hidden Markov chains with values in a finite alphabet, one can show that the penalized maximum likelihood estimator achieves the optimal rate for the exponential decay of the under-estimation probability, see [18].

References

1. J. Rissanen, Modeling by shortest data description. Automatica **14**, 465–471 (1978)
2. A. Barron, J. Rissanen, B. Yu, The minimum description length principle in coding and modeling. IEEE Trans. Inform. Theory **44**, 2743–2760 (1998)
3. P. Massart, Concentration inequalities and model selection, in *Lecture Notes in Mathematics* (Springer, Berlin, 2007). Lectures from the 33rd Summer School on Probability Theory held in Saint-Flour, vol. 1896, July 6–23, 2003 (With a foreword by Jean Picard). ISBN 978-3-540-48497-4; 3-540-48497-3
4. R. van Handel, On the minimal penalty for Markov order estimation. Probab. Theory Rel. Fields **150**, 709–738 (2011)
5. O. Cappé, E. Moulines, T. Rydén, *Inference in Hidden Markov Models* (Springer Series in Statistics, Springer, New York, 2005). With Randal Douc's contributions to Chapter 9 and Christian P. Robert's to Chapters 6, 7 and 13, With Chapter 14 by Gersende Fort, Philippe Soulier and Moulines, and Chapter 15 by Stéphane Boucheron and Élisabeth Gassiat. ISBN 978-0387-40264-2; 0-387-40264-0
6. E. Gassiat, C. Kéribin, The likelihood ratio test for the number of components in a mixture with markov regime, 2000. ESAIM P&S (2000)
7. S. Boucheron, E. Gassiat, Optimal error exponent in hidden Markov model order estimation. IEEE Trans. Inform. Theory **48**, 964–980 (2003)
8. A. Chambaz, A. Garivier, E. Gassiat, A MDL approach to HMM with Poisson and Gaussian emissions. Application to order identification. J. Stat. Plan. Inf. **139**, 962–977 (2009)
9. A. van der Vaart. *Asymptotic Statistics*. Cambridge Series in Statistical and Probabilistic Mathematics, vol. 3 (Cambridge University Press, Cambridge, 1998). ISBN 0-521-49603-9; 0-521-78450-6
10. A. van der Vaart, J.A. Wellner, *Weak Convergence and Empirical Processes* (With applications to statistics. Springer Series in Statistics, Springer, New York, 1996)
11. M. Ledoux, M. Talagrand, Comparison theorems, random geometry and some limit theorems for empirical processes. Ann. Probab. **17**, 596–631 (1989)
12. E. Gassiat, R. van Handel, The local geometry of finite mixtures. Trans. AMS **366**, 1047–1072 (2014)
13. E. Gassiat, R. van Handel, Consistent order estimation and minimal penalties. IEEE Trans. Info. Theory **59**, 1115–1128 (2013)
14. I. Csiszár, P.C. Shields, The consistency of BIC Markov order estimator. Ann. Stat. **28**, 1601–1619 (2000)
15. T. Dumont, Context tree estimation in variable length hidden Markov models. IEEE Trans. Inf. Theory **60**(6), 3196–3208 (2014)
16. E. Gassiat, Likelihood ratio inequalities with applications to various mixtures. Ann. Inst. H. Poincaré Probab. Statist. **38**, 897–906 (2002)

17. J.-M. Azais, E. Gassiat, C. Mercadier, The likelihood ratio test for general mixture models with possibly structural parameter. ESAIM P and S **3**, 301–327 (2009)
18. E. Gassiat, S. Boucheron, Optimal error exponents in hidden Markov model order estimation. IEEE Trans. Info. Theory **48**, 964–980 (2003)

Index

© Springer International Publishing AG, part of Springer Nature 2018
É. Gassiat, *Universal Coding and Order Identification by Model Selection Methods*, Springer Monographs in Mathematics,
https://doi.org/10.1007/978-3-319-96262-7

Printed in the United States
By Bookmasters